Telecommunication

(Volume 1)

Changing Humanities and Smart Application of Digital Technologies

Edited By:

Kuo Hung Huang
Department of E-learning Design and Management,
National Chiayi University,
Taiwan

Bentham Science Publishers Ltd.
Executive Suite Y - 2
PO Box 7917, Saif Zone
Sharjah, U.A.E.
Email: subscriptions@benthamscience.org

BENTHAM
SCIENCE

CONTENTS

FOREWORD

It is with pleasure that I take advantage of the opportunity to write the foreword for this book, which is the work and effort of scholars to explore new ground and participate in research that infuses innovative methods and thinking into academic investigations around people's lives.

Innovative contributions are being made in literature, in history, in geography, in religion, and in education. These contribution are providing the basis for new ways of understanding the contextual nature of technological application practices emerging from humanities research and instruction.

Not only is this book of great value to the digital humanities researchers, but it is equally important and instructive to scholars who are interested in the interdisciplinary research from diverse fields. I hope this book will inspire you to pursue innovative research within existing culture.

Dr. David Tein-Yaw Chung
Department of Computer Science and Engineering
Yuan Ze University
Taiwan

PREFACE

As the title of this book implies, its purpose is to utilize a perspective derived from humanities to consider some of the potential uses of computers in humanities teaching and research. Our intent in producing this book is to provide readers with exemplars of the types of combining computers and subject matters in social science and humanities. Accordingly, our contributors address a variety of computer applications to education and research, which range from humanities, such as literature, geography, history, and religion, to the more technical domains of GIS, mobile devices, and networks.

If this book has any major objective, it is to draw attention to the inter-disciplinary nature of digital humanities, which remain controversial in scholarships or methodologies among academic communities. In doing so, the issues addressed draw on the themes of researching and teaching from a number of subject areas. Our strategy for introducing and synthesizing these issues is to hold an independent and open view of technology adoption. We try to present the various concerns of digital humanities as they emerged in discussions at academic gatherings.

The book is organized loosely into four sections. In Part 1, a chapter by the editor focuses on the situation of digital humanities in scholarly community. Although the advantages are acknowledged prevalently, the debate on applying digital technology on humanities research still continues. The ongoing controversy surrounds the academic values and social acceptance of digital humanities. Thus, more innovative applications in various fields, such as history, religion, or language, should be attempted, reported, or even criticized to inspire elaboration on the forms or values of digital humanities.

In Part 2, four chapters report on adopting spatial technology in humanities research in religion, Chinese literature, geography, and history. GIS (Geographic Information System) is widely used for professional purpose or daily task. With tremendous accumulation of spatial data such as satellite images, street view pictures, and digital maps, GIS provides the humanities researchers with a spatial perspective to explore the connections between places, time, events, or even sentiment. In Chapter 2, the authors introduced the use of Google Earth to conduct an exploratory study on the forbidden religious topic- burial site. In Chinese culture, death is associated with mysteries and taboos. Particularly, conversation about afterlife arrangement with the elderly is seen as an offending behavior. Visualizing the places of good Feng Shui, GIS with 3D views serves as a communication medium for understanding old people's feelings. In Chapter 3, the authors share the innovative instruction experience in a Chinese literature course by combining GIS and famous war in history. Through visualizing maps and animation, students are able to better understand the spatial movements and use of military strategy and tactics while reading paragraphs about a famous war in "Shi-ji". Chapter 4 demonstrates how GIS is capable of researching on history data by analyzing the distribution characteristic of commemorative shrines and steles of the Song Dynasty. After cross-checking the historical documents, these geographic evidences can play a crucial role initiating the cooperation between two fields. The authors of chapter 5 show that GIS transformed history data into drawn maps regarding a historical event, namely Batu Caves Turmoil. And thus, these maps can help to illustrate the cause, the process of and solutions to this event.

In Part 3, three chapters describe how to utilize network applications to support educational activities. Internet usage grows rapidly and tremendously. More importantly, innovative services are developed and incorporated into virtually every aspect of modern life. For example, Internet applications like cloud computing, virtual community, social media, digital library, or MOOCS are creating new forms of social interaction. Also, the online educational materials and activities enable self-learning and encourage new ways of learning, collaboration, and information sharing. Through the cases in chapter 6 to chapter 8, the Internet applications have proven to be a significant influence on education. Chapter 6 introduces the network educational system for Shogi, a Japanese chess game. As an important form of activities to improve skills, post-game discussions frequently take place among Shogi players. The authors designed SAKURA (Shogi Archives and Kansousen Utilities for Research and Advice), an Internet architecture for Shogi. With servers for shared database storing game records and comments, SAKURA supports discussions with shared boards and graphical interfaces visualizing variations of moves. In Chapter 7, the authors describe the process of adopting ODIS (Official Document Information System) in educational administration. Initially, directors and staff resist ODIS to operate on information system because of inconveniences and environmental changes. After constantly interacting with the leadership assignment as leaders or followers, directors and staff gradually accept the new way of distributed leadership. Therefore, work effectiveness is evidently improved. Chapter 8 describes the establishment of eK4 (e-Knowledge Consortium Shikoku) as a local university alliance to provide e-Learning courses related to Shikoku island. For the purpose of facilitating deeper understanding of humanities contents regarding Shikoku region, the consortium organizes e-learning contents of the characteristic lectures at eight universities and implements features of distributed Learning Management System and Shibboleth identification for credit transfer.

In the last part of this book, three chapters examine the effects of the technology on human behavior from a theoretically orientated perspective. Chapter 9 clarifies Transactive Memory Theory that human has the ability to store memory with the aid of the externally stored memories with the advancement of information technology. In chapter 10, the authors present an empirical study indicating that the use of mobile technology as a medium is able to facilitate bilingual language acquisition. Chapter 11 claims the urgent needs for moral development of the Internet users because the anonymity and online free expression significantly characterizes the virtual world which is full of moral dilemmas and value conflicts.

While not attempting to provide a comprehensive examination of all the possible uses of computers in humanities, we purposefully invite contributors to reflect their theoretical philosophies and hand-on experiences, with which we are concerned, as well as a wide range of practical applications. Hopefully, this book will draw the attention of or bring inspiration to digital humanities among academic communities.

Dr. Kuo Hung Huang
Department of E-learning Design and Management
National Chiayi University
Taiwan

List of Contributors

Chen Ming-Chung	Department of Special Education, National Chiayi University, Chiayi, Taiwan
Cheng Ching-Ching	Department of Early Childhood Education, National Chiayi University, Taiwan
Cheng Kuang-Ting	Department of Information Management, National United University, Taiwan
Chuang Su Chong	Graduate Institute of Public Policy, National Chiayi University, Taiwan
Hayashi Toshihiro	Information Technology Center, Kagawa University, Japan
Ho Hsiang-Ju	Department of Early Childhood Education, National Chiayi University, Taiwan
Huang Kuo-Hung	Department of E-learning Design and Management, National Chiayi University, Taiwan
Kinuhata Toshiki	Faculty of Engineering, Kagawa University, Japan
Lee Pei-Lun	Department of E-learning Design and Management, National Chiayi University, Taiwan
Liang Chin-Tu	Graduate Institute of Education, National Chung Cheng University, Taiwan
Liu Hsin-Chun	Department of Education, National Chiayi University, Taiwan
Miura Ryu	Faculty of Engineering, Kagawa University, Japan
Tarumi Hiroyuki	Faculty of Engineering, Kagawa University, Japan
Tsai Jung-Tao	Department of Chinese Literature, National Chiayi University, Taiwan
Yeh Yu-Ching	Department of Early Childhood Education, National Chiayi University, Taiwan

Telecommunication

(Volume 1)

Changing Humanities and Smart Application of Digital Technologies

Telecommunication

Volume # 1

Changing Humanities and Smart Application of Digital Technologies

Editor: Kuo Hung Huang (Taiwan, R.O.C.)

eISBN (Online): 978-1-68108-407-7

ISBN (Print): 978-1-68108-408-4

eISSN (Online): 2542-4580

ISSN (Print): 2542-4564

First published in 2017.

CHAPTER 1

E-research Acceptance and Humanities Community

Kuo-Hung Huang*

Department of E-learning Design and Management, National Chiayi University, Chiayi, Taiwan, R.O.C.

Abstract: Digital humanities is broadly characterized as humanities research associated with data-driven activities. Computational technologies are adopted to explore new methods and tools supplementing established research routines in humanities. However, humanities researchers expressed concerns that the enthusiasm to promote and accelerate the application of technology in humanities teaching and research might overlook the significance of humanities itself. This article points out the advantages of technologies in respond to humanities concerns.

Keywords: Digital humanities, Digital scholar, e-research, Humanities, Mutual dependence, New media, Scholarly communities, Social media, Task-uncertainty, Technology Acceptance.

INTRODUCTION

Digital Humanities, a term used during the early 2000s, was defined, by researchers from a variety of fields, as a new discipline exploring computational methods for traditionally-defined humanities scholarships (Puschmann & Bastos, 2015). In fields of humanities research, such as archive collection, text analysis, map digitalization, and history visualization, computational technologies are adopted to explore new methods and tools supplementing established research routines in humanities (Juola, 2008). Due to its interdisciplinary and technolo-

* **Corresponding author Kuo-Hung Huang:** Department of E-learning Design and Management, National Chiayi University, Chiayi, Taiwan, R.O.C.; Tel/Fax: +886-5-2263411; E-mail: kuohung@mail.ncyu.edu.tw

gical nature, digital humanities are broadly characterized as humanities research associated with data-driven activities. Incorporating a range of different interpretations and innovative technologies, digital humanities has evolved to a unique scholarly field or movement. Studies in media and culture, archive creation and digital library, e-learning, and the prevailing community websites have also been referred to as digital humanities generally (McPherson, 2009).

The suffix "digital", which has affinity to information technology, is progressively used to delimit the new computational areas of humanities research. Due to the pervasive use of communication tools, such as social media, it is most commonly defined as new media, which refers to content accessible on any digital device through interactive user feedback and creative participation on the Internet. With information and communication capabilities to improve intellectual and social life, new media provides a way to interact with other people in a manner that is fundamentally different from what we have had before. People are able to take advantage of expertise spread around the globe or ask virtual friends for advice. Furthermore, the human mind is no longer isolated by location, but is distributed all over the Internet. The entire wealth of stored human knowledge can be accessed through shared mental tools.

CONCERNS FROM THE HUMANITIES COMMUNITY

However, the concept of the use of computers to offer the potential of the expansion of human intellect is little more than a vision, as such claims are not strongly supported by empirical evidences (Olson, 1985; Pea, 1985; Sewell, 1990). Information is not equal to intellect. Access to millions of copies of digital textbooks or online courses of elite universities does not guarantee users' successful learning. Given the availability of such tools and resources, students will rely on finding more and learning less. In fact, the majority of the population use new media for social or entertainment purposes. Furthermore, Striebel (1986) argued that, within a technological framework, all these approaches reflect a shift towards technologizing education, which is consonant with a statement by Elkind (1985) "delegitimize non-technological methods of learning and thinking about problems".

Similar reservations have been expressed by several humanities researchers. They worry that our enthusiasm to promote and accelerate the application of technology in humanities teaching and research might overlook the significance of humanities itself. In spite of the enormous educational value and research inspirations, particularly association with the convenient use of resources, technology is deeply mistaken as a 'fast-food approach' necessarily required in the digital era, and therefore, reduces the educational perspectives and aims that emphasize holistic approaches to nurture humanist whole development.

On the contrary, some humanities researchers in the design field reiterate the importance of digital scholarships. The scholarly digital projects not only provide a researching opportunity for practitioners, but also post an issue regarding role of design in the creation and representation of knowledge for design thinking researchers (Burdick & Willis, 2011). For example, Christine Borgman asserted that, making the digital scholarship an innovating force in humanities research will prevent the community from falling behind in this multimodal and dynamic world, and thus, avoid becoming a victim of the tremendous restructuring of higher education (Borgman, 2009). Additionally, Sams, Lim, and Park (2011) claimed that the online content will provide a rich source of data for academic researches, particularly in social science and humanities. For example, they found that news blogs formed a centerpiece of election campaigns after using blog posts to examine the relationship between the number of votes received by political candidates and the level of their online presence. Consequently, the development of e-Research tool providing access to large online datasets instead of established data collection led to discover new sociopolitical phenomena or theoretical model.

The differences of attitudes regarding information practices between scholarly communities have been revealed by several studies. Digital learning repositories are increasingly important, as they provide resources for teaching, have the potential of fostering innovative new professional behavior, and facilitate the sharing of good pedagogical practices (Millard, Borthwick, Howard, McSweeney, & Hargood, 2013). With abundance of digital collections presently available, integrating primary sources into humanities teaching and learning at the university level becomes more convenient than before; however, collections of online primary sources are still underutilized (Lindquist & Long, 2011). Harley (2007) conducted a study with humanities faculty, and identified several barriers regarding the adoption of digital resources, namely difficulties in finding, maintaining, and selecting appropriate resources for classroom instruction. These issues are not easily resolved without the faculty's devotion of time and awareness of available resources. As Millard, Borthwick, Howard, McSweeney, and Hargood, (2013) claimed, the most significant barrier to participate is innovative behavior itself, as practitioners do not traditionally share their materials or approaches in public.

Another study by Fry (2006) indicated that "the way in which scholarly communities coordinate and control research problems, techniques, strategies, task outcomes, and reputations will significantly influence the production and use of digital infrastructures and resources across fields" (p. 312). Therefore, the author constructed an argument that "patterns the coordination and control within scholarly communities, as shaped by levels of mutual dependence and task-uncertainty, which will influence the development of field-based strategies for information practices within the context of networked digital resources" (p. 312). In other words, with the cultural configuration of knowledge production practices in the digital field, the humanities scholarly community is less concerned with teamwork on producing field-based digital resources and integrating them into their epistemology.

LOOKING FORWARD

Meaning is created between people as their viewpoints are expressed in discourse. The challenge for digital humanities researchers is creating a shared social reality through continuous endeavors that give insights into the criteria and facilitate reflection on meaning making process. Soon and Park (2009), from the social science perspective in an Asian context, asserted that main obstacles preventing social scientists and humanities from approaching and embracing new technologies to conduct research work was lack of technical knowledge and skills. In addition to education and training programs to enhance social scientists' expertise combining computational science and social sciences, a practical change to overcome the barrier will be promoting collaborative efforts among social scientists, computer professionals, and humanities scholars.

A number of digital humanities projects have indicated that the success of digital humanities depends on selecting the appropriate technology. For example, a study on digitizing an old text verifies that the technological tools appropriate for the task can facilitate accuracy as well as critical reflection, and thus, avoid the alleged risk of shallowness associated with utilizing computers (Carr, 2010; Delfino, 2011). Led by technological change, new challenges and opportunities will motivate the development of new dimensions for humanities, and eventually evolve new digital scholarships. The sense of urgency has been articulated clearly- the humanities research field is in great need of innovation.

CONFLICT OF INTEREST

The author confirms that the author has no conflict of interest to declare for this publication.

ACKNOWLEDGEMENTS

The author would like to thank the Ministry of Science and Technology for the financial support of this project (NSC 102-2420-H-415-004-MY2).

REFERENCES

Borgman, C. L. (2009). The digital future is now:a call to action for the humanities. *Digital Humanities Quarterly, 3*(6) online serial.

Burdick, A., Willis, H. (2011). Digital learning, digital scholarship and design thinking. *Design Studies, 32*(6), 546-556.
[http://dx.doi.org/10.1016/j.destud.2011.07.005]

Carr, N. (2010). *The shallows: What the Internet is doing to our brains.*. New York: Norton & Co..

Delfino, M. (2011). Against BibliOblivion: How modern scribes digitized an old book. *Computers and Education, 57*(3), 2145-2155.
[http://dx.doi.org/10.1016/j.compedu.2011.05.018]

Elkind, D. (1985). The impact of computer use on cognitive development in young children: A theoretical analysis. *Computers in Human Behavior, 1*(2), 131-141.
[http://dx.doi.org/10.1016/0747-5632(85)90026-3]

Fry, J. (2006). Scholarly research and information practices: A domain analytic approach. *Information Processing and Management,* (42), 299-316.

[http://dx.doi.org/10.1016/j.ipm.2004.09.004]

Harley, D. (2007). Use and users of digital resources: a survey explores scholars' attitudes about educational technology environments in the humanities. *EDUCAUSE Q., 30*(4), 12-20.

Juola, P. (2008). Killer applications in digital humanities. *Literary and Linguistic Computing, 23*(1), 73-83. [http://dx.doi.org/10.1093/llc/fqm042]

Lindquist, T., Long, H. (2011). How can educational technology facilitate student engagement with online primary sources? A user needs assessment. *Library Hi Tech, 29*(2), 224-241. [http://dx.doi.org/10.1108/07378831111138152]

McPherson, T. (2009). Introduction: Media studies and the digital humanities. *Cinema Journal, 48*(2), 119-123.

Millard, D.E., Borthwick, K., Howard, Y., McSweeney, P., Hargood, C. (2013). The HumBox: Changing educational practice around a learning resource repository. *Computers and Education,* (69), 287-302. [http://dx.doi.org/10.1016/j.compedu.2013.07.028]

Olson, D.R. (1985). Computers as tools of the intellect. *Educational Researcher, 14*(5), 5-8. [http://dx.doi.org/10.3102/0013189X014005005]

Pea, R.D. (1985). Integrating human and computer intelligence. In: Klein, E.L., (Ed.), *Children and Computers: New directions for child development.* (pp. 75-96). San Francisco: Jossey Bass.

Puschmann, C., Bastos, M. (2015). How digital are the Digital Humanities? An analysis of two scholarly blogging platforms. *PLoS One, 10*(2), e0115035. [http://dx.doi.org/10.1371/journal.pone.0115035] [PMID: 25675441]

Sams, S., Lim, Y.S., Park, H.W. (2011). E-research applications for tracking online socio-political capital in the Asia-Pacific region. *Asian Journal of Communication, 21*(5), 450-466. [http://dx.doi.org/10.1080/01292986.2011.594897]

Sewell, D.F. (1990). *New tools for new minds: A cognitive perspective on the use of computers with young children.*. London: Harvester Wheatsheaf.

Soon, C., Park, H.W. (2009). The rise of e-science in Asia: Dreams and realities for social science research. Case studies of Singapore and South Korea. In: Jankowski, N., (Ed.), *E-research: Transformations in Scholarly Practice.* (pp. 109-128). London: Routledge.

Striebel, M.J. (1986). A critical analysis of the use of computers in education. *Educational Communication and Technology, 34*(5), 137-161.

CHAPTER 2

Feng Shui and Technology: Case Studies of Spatial Information Applications

Kuo-Hung Huang[1,*] and **Ching-Ching Cheng**[2]

[1] *Department of E-learning Design and Management, National Chiayi University, Chiayi, Taiwan, R.O.C.*

[2] *Department of Early Childhood Education, National Chiayi University, Chiayi, Taiwan, R.O.C.*

Abstract: This paper reports several case studies on creating innovative applications of spatial technology with Feng Shui in order to meet the requirements of burials. Case study research is used to investigate how individual behaviors are influenced by integrating technology with Feng Shui, within its context. A qualitative design is chosen to collect data from interviews, participatory observations, and demonstrations. The researcher interviewed the case subjects in order to assess their needs, and then accompanied them on their trips to family tombs and recorded the traverse path. Finally, this researcher displays the maps and discusses the applications. Findings indicated that Google Earth could be used with Feng Shui for filtering information regarding site selection, as well as recording and sharing information on burial activities. Above all, spatial information on the Internet could serve as a medium for cultural communications between generations. This study increases our understanding of the meaning of Feng Shui to our elders, and how to integrate spatial information with traditional culture.

Keywords: Burial, Burial site, Cultural communication, Elderly, Feng Shui, Google Earth, Spatial information, Traditional culture.

INTRODUCTION

Originally developed in China, the Feng Shui theory has long been used in architecture and landscape design and planning in East Asian countries, and has

* **Corresponding author Kuo-Hung Huang:** Department of E-learning Design and Management, National Chiayi University, Chiayi, Taiwan, R.O.C; Tel/Fax: +886-5-2263411; E-mail:kuohung@mail.ncyu.edu.tw

been extending its influences into daily lives (Huang & Teng, 2009; Mak & Ng, 2005). The main premise of the Feng Shui theory is to maintain a harmonious relationship between humans and environment. The literal meaning of Feng Shui is wind and water, and its major concerns are protecting sites from wind, and obtaining fresh water. The theoretical basis of Feng Shui relies on the concept of chi, the vital force that influences the fate of the space's occupants. Accordingly, decision-making regarding living spaces and burial sites are at the center of Feng Shui practices (Chang & Lii, 2010; Chen & Wu, 2009). In fact, Feng Shui has developed into two main fields, namely, "Yang House" Feng Shui, which is applied to palaces, cities, villages, and housing, and "Yin House" Feng Shui, which is applied to graves. Both have similar criteria for deciding grave and housing locations. Believing the quality of their Feng Shui influences both the physical and mental health of the inhabitants, many Chinese link the construction and the main direction of their houses with their fate (Mak & Ng, 2005).

The Chinese believe that fortune and health follow the gathering of positive chi. According to Feng Shui principles, the site of human dwellings must be located at a place where the heavenly chi and earthly chi are constantly interacting in harmony (Chang & Lii, 2010). Feng Shui concerns the physical form of the site under consideration and its surrounding environment. For an auspicious orientation, the Feng Shui master searches for propitious locations and directions, such as a favorable mountain range, lucky surrounding hills, or a nearby river. The process of searching is unpredictable, time consuming, and sometimes disturbing, particularly if a location is in urgent need, as in cases of funerals. Field observations have the disadvantage of only providing limited information on the area-wide distribution of the Feng Shui locations.

In China and Korea, burying one's parents in a place deemed propitious under Feng Shui principles is seen as a way to demonstrate filial piety. Most individuals take the responsibilities of maintaining and cleaning their ancestor's burial areas. Therefore, Feng Shui practice is the embodiment of cultural beliefs in burial activities. In recent years, spatial technology has become prevalent and is easily used. These emergent resources can be used as tools for the practice of Feng Shui. In this paper, the researcher reports several case studies on the Feng Shui of tombs by integrating Global Positioning System (GPS) and Google Earth capabilities, and creates an innovative application of spatial technology through Feng Shui in order to meet the needs of individuals. Through interactive demonstrations and communications, the case studies can enhance our understanding of the contextual meaning of Feng Shui for the elderly.

FENG SHUI AND BURIAL SITE SELECTION

The Feng Shui of tomb location is extremely important in East Asia, where it is a common belief accepted by the public that the location of the tomb would affect the descendant's well-being (Um, 2009). In addition, it is well known that ancient Chinese emperors consulted Feng Shui experts before building their tombs. Although many consider Feng Shui as a superstition, the spirit of Feng Shui is still tightly integrated into individual lives. The mysterious forces of Feng Shui on tombs are commonly believed to be associated with health, prosperity, and good luck. For example, several leaders of major political parties in South Korea have moved their ancestors' graves to propitious locations. In Taiwan, Chiang Kai-shek's rise to power is said to be based on the good Feng Shui of his mother's grave. His downfall is blamed on the communists later relocating this grave (Rossbach, 1983). Even today, some TV programs predict who will win the presidential election by comparing the Feng Shui of candidates' ancestral tombs.

The direction and position of a grave are carefully selected by Feng Shui experts who concentrate on the analysis of sites. They first observe the land formation and terrain, and then determine location and orientation. An ideal Feng Shui site for a tomb is backing the tomb onto a mountain side and facing the water side (Hwangbo, 2002; Tam, Tso, & Lam, 1999). A watercourse is supposed to flow from an auspicious direction towards an inauspicious direction, which symbolizes sweeping bad things away, and the site is enclosed by mountains with an opening towards the south. The site chosen by this Feng Shui method is supposed to be the place where chi from Heaven and Earth meet, and is therefore symbolically centered in the cosmos. The Feng Shui spot, "hsueh", is the very place where chi is clustered, thus a house or a tomb is erected, and the courtyard in front of the location is called ming-tang, literally meaning "a bright house" where chi comes through (Hwangbo, 2002).

Despite the fact that there is no scientific evidence substantiating its influence on people, Feng Shui has been so successful in application as its guidelines are often straightforward and clear, offering quick and easy solutions to solve dilemmas (Bonaiuto, Bilotta, & Stolfa, 2010; Xu, 1998). Furthermore, during the process of decision-making for dwellings or moving, Feng Shui may provide individuals with an additional source of information, thus, reducing any uncertainty-induced anxieties (Tsang, 2004a, 2004b).

SPATIAL TECHNOLOGY

Although area-wide data, such as a topographic map of villages and towns sites in landform Feng Shui have been used, it is rare to apply technology in finding locations that are fitting as important spatial variables for Feng Shui practice in large areas (Mak & Ng, 2005; Um, 2009). Geographic Information Systems (GIS) are a tool for the collection, storage, manipulation, analysis, and display of data, and is commonly used to support local fieldwork activities by providing maps and terrain information (Balram & Dragicevic, 2008; Slocum, 1999; Warburton & Higgitt, 1997). In recent years, Web-based GIS systems, such as Google Earth, promote the utilization of spatial information in daily lives. In this study, two spatial technologies are used, namely Google Earth and GPS. Google Earth provides on-line data resources, such as a digital atlas, satellite images, photos, and streaming media. With interfaces designed to increase the level of interactivity, mapping can be more dynamic by providing users with information related to a place, and thus, enhancing the user's experience as a novice explorer (Patterson, 2007). For example, the user can see the Grand Canyon from space, and even from various angles in order to appreciate its depth. In addition, Google Earth has powerful functionalities, such as a virtual flythrough and 3D visualization, which require significantly less time on user training. GPS is a space-based radio system that freely provides reliable positioning, navigation, and timing services to the public. Equipped with the GPS receiver, users can accurately locate where they are by retrieving the information on location and time that is continuously transmitted from the satellite. Today, more mobile devices are GPS attached, providing navigation for transportation, crime prevention, and other location-based services.

In this study, a watch-like GPS tracker is used to integrate with Google Earth, which continually stores time and location information in built-in memory that can be easily downloaded and converted to Google Earth kmz data files. Once a user opens a received kmz file, the Google Earth system will automatically display the time, labels, tracks, and comments for a satellite image.

METHODS

Especially appropriate for new topic areas, the aim of case study research is to investigate phenomena within its context in order to understand how behavior or processes are influenced, as well as the influence context (Eisenhardt, 1989). A case study is not a method but a strategy, and generally includes multiple methods to flexibly probe a complex phenomena (Harley, 2004). In this study, a qualitative design was chosen to provide a chain of evidence contributing to the comprehension of the case study conclusions (Yin, 1994). Data were collected through several stages in order to provide the basis for understanding the interactive and evolving processes of technology adoption. Depending on the schedule of each case, the time interval between stages is varied. The following are the stages and data sources of this study:

- Stage 1: Interviews - Individual in-depth interviews were conducted with each of the case subjects. The researcher interviewed the case subjects to assess their needs and informal conversations provide additional data.
- Stage 2: Participatory observation - The researcher joined tomb related activities in order to collect data. A GPS tracker was carried to record the traverse path.
- Stage 3: Demonstrations - The researcher displayed the maps created with the GPS data collected in the previous stage. Interactive demonstrations, informal conversations, and focus group were conducted to collect data.

CASE OF MR. HUANG

Mr. Huang's father, who passed away in 1982, was a successful business man. His families firmly believe in Feng Shui and considered a good location for a tomb which would benefit the descendants of the tomb's occupant. "It was a hard task", Mr. Huang recalled the painful experiences of finding a suitable grave, "My father suddenly passed away and I was still in a gloomy mood. Nevertheless, I had to search for a burial site as soon as possible. Everyone in the family asked me the result each time I returned home. I almost collapsed facing the tremendous pressure. The funeral was not held until nearly a month later when I found a good burial site". Mr. Huang described the inefficient process, "I, along with the Feng Shui master, drove to evaluate the potential locations referred by friends. After numerous times of exploration, we finally spotted a perfect location". As shown in Fig. (**1**), the targeted location sits on a hill and faces two ponds. In Feng Shui, this tomb would be a blessing to the offspring of the buried; however, this area was prohibited from burying by law. After rounds of negotiations, they reached an agreement with the land owner, where they were privately allowed to bury at the location, but promised to keep the deal a secret. For decades, the Huang's family believed that the tomb of Feng Shui had brought them good luck because the 3 grandsons of the buried became a professor, a medical doctor, and a successful business man.

Fig. (1). A screen shot of Mr. Huang's tomb from Google Earth.

Mr. Huang also believed that a damaged or abandoned burial site would bring inauspiciousness. Several years ago, a flood brushed away the foundation of the tomb, and as a result, he spent a great deal of money fixing it. However, he is getting old and expects his son and nephews to put more concern into this family matter. Nobody but Mr. Huang knew the exact position of the tomb, and he wanted to pass the information to the younger generation in order that they could periodically visit their ancestor's tomb. In addition, he hoped that the tomb sweeping activities could be recorded and shared by other family members in order to tighten the bonds of the Huang family.

The researcher was invited to join a family trip to the tomb. With the GPS tracker, the path taken to tomb sweeping activities was recorded and uploaded to Google Earth on the Internet, the path and exact position were also packaged as a file and distributed to his family (Fig. **2**). Mr. Huang was excited to see the tomb on a satellite image. He was more surprised by the 3D representation in Google Earth (Fig. **3**), which can be used to visualize the terrain and facilitate searching for location of good Feng Shui.

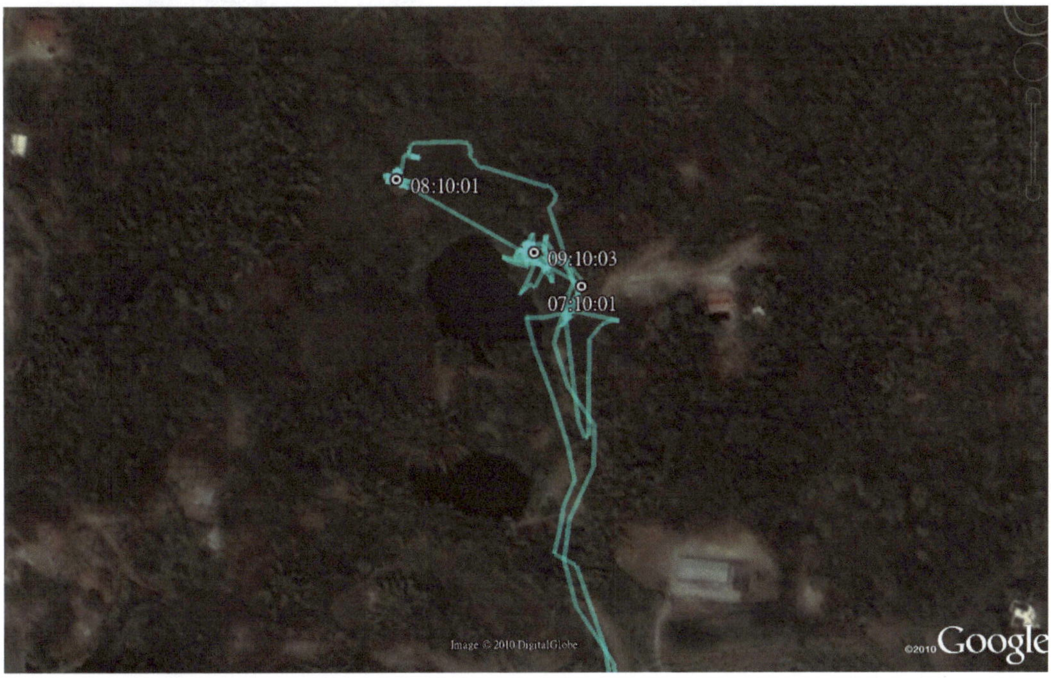

Fig. (2). A screen shot of the path to visit Mr. Huang's tomb from Google Earth.

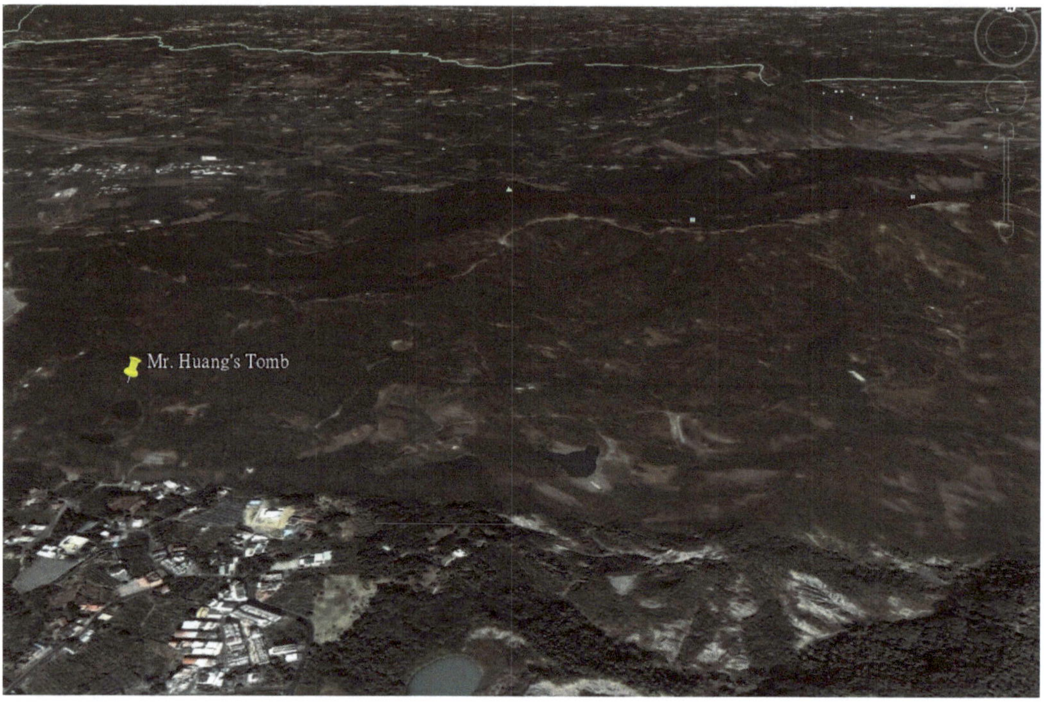

Fig. (3). A 3D-view screen shot of Mr. Huang's tomb from Google Earth.

CASE OF MRS. WANG

Due to decreasing undeveloped land and stricter burial laws, urban residents have chosen cremation instead of burial. In addition to accepting the idea of cremation, many even select the cremation urn site while they are alive in order to save trouble for their families. Mrs. Wang is a 90-year-old Kaohsiung city resident. She is a devoted Buddhist and Feng Shui advocator, and although she decided on cremation after passing away, Mrs. Wang is still convinced that the placement of her cremation urn would influence the fate of her offspring. As the Feng Shui principle suggests, the cremation urn pagoda should sit on a hill and face water. Furthermore, more temples are being constructed in affiliated buildings to accommodate cremation urns. It is commonly believed that the soul can hear the monks' chanting and be sheltered by the Buddha. After surveying several cremation urn pagodas in the Kaohsiung metropolitan area, Mrs. Wang and her family eventually chose a pagoda in the Yuan Heng temple. As Fig. (**4**) shows, the Yuan Heng Temple sits on the Chaisan Mountain and faces the Love River.

However, all the urn spaces in the temple were sold. In fact, most people purchase them in advance for future needs, while some choose it as an investment, particularly those urn spaces with good Feng Shui. Through brokers, Mrs. Wang later obtained an urn space in the temple. Mrs. Wang hopes all members of her family know the pagoda's location in order to worship her in the future. This researcher used a GPS tracker to record the path from the Yuan Heng Temple to the rail station, and then packaged the information as a kmz file to distribute to her family members. The researcher had successfully used Google Earth software to meet Mrs. Wang's needs. Additionally, she is interested in a map with all urn pagodas in the city that she may refer to her friends (Fig. **5**).

Fig. (4). A screen shot of Yuan Heng Temple from Google Earth.

Fig. (5). A screen shot of urn pagodas and temples in Kaohsiung city from Google Earth.

CASE OF MRS. LING

Mr. Ling was a rich farmer who owned vast amounts of land in the coastal areas. Twenty years ago he died in a car accident. Traditionally, farmers were buried in their own farmland. Mr. Ling was buried adjacent to his father's tomb on the family farm. His tomb faced an irrigation trench in the south, and although there was not any hill around this area, Mr. Ling's family believed that the Feng Shui of their ancestor's tombs would bring good fortune to their offspring because all Mr. Ling's sons were blessed with great achievements, as both of Mr. Ling's sons earned a Ph.D. from George Tech, and one worked as the Vice President of Merrill Lynch, while the other became a researcher in Boeing.

Mrs. Ling, at nearly 90 years old, walks to visit the tombs once a week. She proposed to hold a ceremony of "bone picking", a process of moving the buried body to a cremation urn, but could not reach a consensus with all family members. In Chinese society, some people will conduct bone picking after the body has been buried for over seven years, and the main reasons for holding bone picking include land development or problems in tomb maintenance. Before the process, Feng Shui experts pick an appropriate day and time to exhume the grave, and then the bone pickers cremate the body of the dead and collect the bones into an urn. Later in the day, the urn is moved to an urn pagoda, and the original tomb is then destroyed. Since her sons seldom return to Taiwan and the grown overseas grandchildren are alien to the Chinese tradition, she worries that the farms could be sold after she dies and the tombs on the farm would be inconsiderately removed, which would eventually lead to bad fortunes for the offspring. To help the young generation to remember the location of the tombs and the clan cremation urn house, she walked the researcher to these locations for data collection. As shown in Fig. (**6**), Google Earth is used to distribute the information to her families in the U.S. Mrs. Ling was not sure if the information was helpful, but she appreciated the work, which clearly identifies the locations.

Fig. (6). A screen shot of Mr. Ling's tombs and the clan urn house from Google Earth.

CONCLUSION

Feng Shui is an important element of Chinese superstitious beliefs that affect daily lives. Although what actual benefits Feng Shui brings are still uncertain, Feng Shui, taken as accumulated culture and spiritual support, continues to be influential. For instance, despite being constantly criticized and censured as a superstition by a communist government, Feng Shui practice continues to rage in China. Particularly in the context of insecurities created by the new market economy, summoning good fortune has renewed its purchase on the activities of daily life (Knapp, 1999). As Hwangbo (2002) claimed, "Feng Shui is a trace of the development of human consciousness" (p. 126), and the erudition, in terms of art and philosophy, can nourish and enrich our late modernity.

Summarized from the above case studies, the concerns of the elderly regarding Feng Shui were oriented to sustain a bright future for their families. Technology plays the role of information provider and communication facilitator to maintain the cultural ties between families. In practical use, Google Earth can be used with Feng Shui practices in two ways, first, spatial technology helps to filter information for site selection. With satellite images, individuals can locate water features, such as ponds or creeks. In addition, the current version of Earth Google provides elevation data used to identify hilly areas. The 3D data visualization function can facilitate the process of identifying favorable spots in mountainous areas. The other capability is to record and dispatch activity information. In China and Korea, tomb sweeping day is a national holiday, and stands for an important event for family members to gather together (Um, 2009). Maintaining and worshiping the tombs of their ancestor is not simply a routine activity, but is regarded as a way of sustaining inheritance to family values and clan bonds. Along with GPS, the path of visiting specific locations can be appended on a map and reserved as a digital memory. Through technology, the course of the activities can be archived and shared by family members beyond time and distance limitations.

This study is intended as an initial investigation regarding the utility of Google Earth in Feng Shui practices. Through exploration, we may come nearer to understanding the meaning of Feng Shui to our elderly, and how we can integrate spatial information with traditional culture. In East Asia, younger generations have lost the keen sense of understanding the traditional meanings of symbolic folk culture, and Feng Shui plays the role of a chain of life that connects an individual with both ancestors and descendants (Knapp, 1999; Yoon, 2006). Spatial information on the Internet not only serves as a tool, but also as a medium for cultural communication between generations.

As noted by Downey (2007), there continues to be a need to research the flows of spaces, as well as the social shaping of space. Current thinking in human geography, especially notions of processes, contexts, and relationality, should be of interest to information sciences in the research community. From this study, it is hoped that information technology, and its impacts on Feng Shui, will bring additional insights to the sustainability and innovative approaches of information sharing of family traditions. A further implication for future research includes developing a geographic-based Feng Shui information system on the Internet and the critically-evaluated measurements to understand its influences.

CONFLICT OF INTEREST

The authors confirm that the authors have no conflict of interest to declare for this publication.

ACKNOWLEDGEMENTS

The authors would like to thank the Ministry of Science and Technology for the financial supports on the project (NSC 102-2420-H-415-004-MY2).

REFERENCES

Balram, S., Dragicevic, S. (2008). Collaborative spaces for GIS-based multimedia cartography in blended environments. *Computers and Education, 50*(1), 371-385. [http://dx.doi.org/10.1016/j.compedu.2006.07.004]

Bonaiuto, M., Bilotta, E., Stolfa, A. (2010). Feng Shui and environmental psychology: A critical comparison. *Journal of Architectural and Planning Research, 27*(1), 23-34.

Chang, W.L., Lii, P. (2010). Feng Shui and its role in corporate image and reputation: A review from business and cultural perspectives. *Journal of Architectural and Planning Research, 27*(1), 1-14.

Chen, X.Q., Wu, J.G. (2009). Sustainable landscape architecture: Implications of the Chinese philosophy of “unity of man with nature” and beyond. *Landscape Ecology, 24*(8), 1015-1026. [http://dx.doi.org/10.1007/s10980-009-9350-z]

Downey, G. (2007). Human geography and information studies. *Annual Review of Information Science*

Technology, 41, 683-727.
[http://dx.doi.org/10.1002/aris.2007.1440410122]

Eisenhardt, K.M. (1989). Building theories from case study research. *The Academy of Management Review, 14*(4), 532-550.

Harley, J. (2004). Case study research. In: Cassell, C., Symon, G., (Eds.), *Essential guide to qualitative methods in organizational research.* (pp. 323-333). London: SAGE.
[http://dx.doi.org/10.4135/9781446280119.n26]

Huang, L.S., Teng, C.I. (2009). Development of a Chinese superstitious belief scale. *Psychological Reports, 104*(3), 807-819.
[http://dx.doi.org/10.2466/PR0.104.3.807-819] [PMID: 19708408]

Hwangbo, A.B. (2002). An alternative tradition in architecture: Conceptions in Feng Shui and its continuous tradition. *Journal of Architectural and Planning Research, 19*(2), 110-130.

Knapp, R.G. (1999). *China's living houses: Folk beliefs, symbols, and household ornamentation.* Hawaii: University of Hawaii Press.

Mak, M.Y., Ng, S.T. (2005). The art and science of Feng Shui - a study on architects' perceptions. *Building and Environment, 40*(3), 427-434.
[http://dx.doi.org/10.1016/j.buildenv.2004.07.016]

Patterson, T.C. (2007). Google Earth as a (not just) geography educational tool. *Journal of Geography, 106*(4), 145-152.
[http://dx.doi.org/10.1080/00221340701678032]

Rossbach, S. (1983). *Feng Shui: The Chinese art of placement.* New York: Penguin Books.

Slocum, T. (1999). *Thematic Cartography and Visualization.* Upper Saddle River, NJ: Prentice Hall.

Tam, C.M., Tso, T.Y., Lam, K.C. (1999). Feng Shui and its impacts on land and property developments. *Journal of Urban Planning and Development, 125*(4), 152-163.
[http://dx.doi.org/10.1061/(ASCE)0733-9488(1999)125:4(152)]

Tsang, E.W. (2004). Superstition and decision-making: Contradiction or complement? *The Academy of Management Executive, 18*, 92-104. a
[http://dx.doi.org/10.5465/AME.2004.15268696]

Tsang, E.W. (2004). Toward a scientific inquiry into superstitious business decision-making. *Organization Studies, 25*, 923-946. b
[http://dx.doi.org/10.1177/0170840604042405]

Um, J.S. (2009). Exploring spatially prioritized parameters of Feng-Shui from tomb footprints. *International Journal of Geographical Information Science, 23*(4), 513-529.
[http://dx.doi.org/10.1080/13658810802055954]

Warburton, J., Higgitt, M. (1997). Improving the preparation for fieldwork with IT: Two examples from physical geography. *Journal of Geography in Higher Education, 21*(3), 333-348. [http://dx.doi.org/10.1080/03098269708725440]

Xu, P. (1998). 'Feng-Shui' models: Structured traditional Beijing courtyard houses. *Journal of Architectural and Planning Research, 4*, 271-282.

Yin, R. (1994). *Case Study Research Design.* Thousand Oaks, CA: SAGE.

Yoon, H-K. (2006). *The culture of Feng Shui in Korea: An exploration of East Asian geomancy.* Plymouth, UK: Lexington Books.

CHAPTER 3

Application of GIS in the Teaching of “Shi-Ji“- <Huaiyin Hou Biography> as an Example

Jung-Tao Tsai[1] and **Pei-Lun Lee**[2,*]

[1] *Department of Chinese Literature, Chiayi University, Chia-Yi, Taiwan, R.O.C.*

[2] *Department of E-Learning Design and Management, National Chiayi University, Chia-Yi, Taiwan, R.O.C.*

Abstract: This article was set out to be a new attempt at the teaching and research of “Shi-ji”. It was supposed to create maps for the sections and chapters in “Shi-ji” relating to powerful generals with the assistance of GIS so that the readers could better understand the spatial movements and use of military strategy and tactics while reading paragraphs about war in “Shi-ji”. The method won acclaims from most students in the preliminary classroom experiment, who showed the wish that more maps be created in the future to improve the teaching and research of “Shi-Ji”.

Keywords: Biographies, Generals, GIS, Google earth, Han dynasty, Han Xin, Huaiyinhou, Military strategy, Shi-Ji, Spatial movements.

INTRODUCTION

“Shi-Ji” was written by Sima Qian in the second century B.C. as the first biographical general history book in China, imposing profound influences on history, literature, or even intellectual thinking of later generations. It is unashamedly a classic of Chinese culture and reading “Shi-Ji” has always been the common experience of traditional scholar-officials. To this day, most departments of Chinese studies and departments of history are still offering courses relating to “Shi-Ji”. When Fu Sinian was the Chancellor of the National

* **Corresponding author Pei-Lun Lee:** Department of E-Learning Design and Management, National Chiayi University, Chia-Yi, Taiwan, R.O.C; Tel: 886-5-2717226; Fax: 886-5-2062328; E-mail: peilun@mail.ncyu.edu.tw

Kuo Hung Huang (Ed.)

Taiwan University, all the freshman students were required to read "Shi-Ji" closely. The two texts of "Shi-Ji", Zhangshizhi Zhifa and Hongmenyan, are included respectively in the middle school and senior high school Chinese textbooks. Of all the monographs concerning "Shi-Ji", two thirds are the attempts to popularize the work, including abridged editions, annotated editions, translated editions in vernacular Chinese and adaptations. It is thus clear that neither school students, nor the general public are unfamiliar with "Shi-Ji", but rather interested in it.

"Shi-Ji" is not only widely known and spread in Chinese cultural circles, but also fruitfully studied in international sinology. It was introduced to Japan in the Tang Dynasty (8th century A.D.) and to Korea before that. It has complete translations (containing the130 volumes of "Shi-Ji") in Japanese, Korean and English and the complete translation of it in Russian has also been published. "Shi-Ji" could thus well be called the precursor in the spreading of Chinese culture. When invited to deliver a paper on "Shi-Ji" on the international conference on sinology of the European Association for Chinese Studies in 2014, the author found that many scholars from Russia, America and Hong Kong had read "Shi-Ji" closely and offered relevant courses. "Shi-Ji" could thus well be a common language in sinology, whose significance will be increasingly obvious in the course of globalization.

The reading of classics begins with the study of words, followed by in-depth reading and hard thinking to discern the messages of the classics, and ends in the dialogue between oneself and the classics, when the classics will be activated and their contemporary significance be brought out. In this exploration of the classic's contemporary significance, words are the bridge while the dialogue with oneself is the central pathway. In the reading of historical books, apart from the mastery of words and temporary sequence, the spatial background is also indispensable. This is especially true when it comes to the narratives about powerful generals. If the spatial contextual illustrations could be added in the picture, it will surely be very conducive to the teaching of "Shi-Ji".

This paper was thus set out to explore the new possibilities of teaching "Shi-Ji" by studying "Shi-Ji" and the relevant texts about powerful generals with the help of

GIS. It was hoped that through this interdisciplinary effort, the interpretation of "Shi-Ji" could be spatialized and contextualized, the teaching content of "Shi-Ji" could be deepened and enriched, and the students be more understanding of the classic wisdom, hence elevating their life. This paper is comprised mainly of three parts: The first part is devoted to the war narrative in "Shi-Ji"-Huaiyinhou Biography, followed by the second part describing the application of GIS in the teaching and research of "Shi-Ji", with "Shi-Ji"-Huaiyinhou Biography as the example. The feedbacks from the students on the GIS-assisted "Shi-Ji" teaching are analyzed in the last part, which also points out the room for future studies.

ANALYSIS OF THE THREE BATTLES IN HUAIYINHOU BIOGRAPHY

When reading "Shi-Ji", especially the biographies, the readers could easily notice the different number of characters that are depicted in each biography. Some of them might be devoted solely to one person, such as Wuzixu Biography. Others might be the combined biography of two people, like Guan Yan Biography. Still others might be the classified biography of over three people, namely the total of 10 chapters including Cike Biography, Xunli Biography and Huozhi Biography. Dealing with the history of more than 2,500 years in 130 chapters and more than 500,000 words, it is necessary for "Shi-Ji" to have its own system and framework to address the chronological span. The framework is constituted by the five categories of Basic Annals (benji), Tables (biao), Treatises (shu), Hereditary Houses (shijia) and Ranked Biographies (Biography) while the characters are classified in the feasible manner of handling complexity by simplicity. The classified biographies are designed to demonstrate the historical development *via* several categories of characters. These characters, therefore, should be representative in the historical development, besides featured by clustering effects. Take Kuli Biography for example. It depicts a group of officials favored by Emperor Wu that knowingly break and interfere with the law. It thus not only reflects the bureaucratic ecology, but also satirizes Emperor Wu's rule by draconian law. Other important categories of characters, on the other hand, are not presented in the form of classified biographies. Rulers, for example, are mostly dedicated with an entire chapter like Gao Zu Benji due to their prominent status and profound influence. Basic annals could thus be treated as a "large classified biography" comprised of 12 chapters. This is also true for Hereditary Houses.

Among the biographies, powerful generals are an important character type apart from those described in the 10 chapters of classified biographies. For the purpose of heading the troops to behead the enemy general and take down the enemy's flag, they have to be proficient with combat skills and fight as grim warriors themselves. The most important thing for these powerful generals is to kill the enemy and send the victory. That's why they have to be equipped with rich knowledge of military strategy and tactics and use them freely. Since these powerful generals are often honest and straightforward in personality and placed in the battlefield all year along, given their high position, they are usually closely watched. They have thus to be highly politically intelligent to survive in the complex officialdom. The art of war being part of Sima Qian's family education, the Grand Historian was deeply sympathetic with the violent ups and downs in the lives of powerful generals. He thus devoted about 20 chapters of “Shi-Ji” solely to over 30 powerful generals, covering more than 17% of the work. Among them, Huaiyinhou Biography, which is dedicated to Han Xin, one of the three heroes of the early Han dynasty, is a famous chapter on powerful generals, as well as a frequently selected piece in the teaching of “Shi-Ji”. In the works of Mao Kun in the Ming Dynasty, Han Xin was honored as the Fairy Strategist and his art of war was included among the military strategy and tactics in Hanshu Yiwenzhi Zhuzilue (a chapter listing and describing military works), namely the top of all military works. These are all proof that Han Xin was a top general that could maneuver troops like a god. Sima Qian dedicated almost one third of Huaiyinhou Biography to the four battles led by Han Xin. The surrender of the Yan State was achieved through threat of force and diplomatic means without deploying any troops and thus a sublime expression of the tactics of “to subdue the enemy without fighting” in the Art of War. Given exactly that there was no fighting involved, it does not take much space in the chapter. The war narrative here is therefore focused on the three battles of conquering Wei, breaking Zhao and defeating Qi.

Conquering Wei

In the second year of Han (205B.C.), Liu Bang marched out of Hanzhong and seized control of Kuan Chung, after which he went east from the Hangu Pass and broke Pengcheng with an army of 560,000 people, winning a significant victory in the Competition between Chu and Han for Hegemony. When the joint forces of Liu Band was celebrating for the victory, however, Xiang Yu went back from Qi with 30,000 picked troops, putting the troops of Liu Bang in rout instantly and leaving Liu Bang fleeing to the west with less than three hundred remaining forces. The kings and lords, who had pledged allegiance of Han previously, claimed independence again one after another or turned to Xiang Yu instead. Among them, Bao, king of Wei asked to be allowed to go home and look after his family. When he reached Wei, however, he cut off the fords over the Yellow River, which was deemed as a revolt against Han. Liu Bang thus dispatched Han Xin to attack Wei. Han Xin adopted the tactics of making a feint to the east but attacking in the west. He planted a great army at Linjin, a ferry on the Yellow River, making the false impression that he would attack Wei from here. The king of Wei believed him and concentrated his forces at Puban to wait for their arrival. In the meantime, Han Xin was secretly leading forces to the north and ferried across the river in wooden tubs by way of Xia Yang. Taking the forces of Wei completely by surprise, the Han troops broke Anyi, the capital of Wei and took the king of Wei as prisoner. Shortly after Han Xin's victory, Liu Bang took command of his best troops and brought them to Xingyang to help in the blockage against Xiang Yu.

Breaking Zhao

In the third year of Han (204B.C.), Han Xin and Zhang Er led a force of tens of thousands to attack Zhao. Chen Yu, a general of Zhao, gathered an army numbering 200,000 at the mouth of Jing Gorge to wait for the arrival of the enemy. Jing Gorge was the only way to Zhao featuring a road so narrow and difficult that two carts cannot drive side by side, nor two horsemen ride abreast. It could thus be well called impregnable. Li Zuoche, a tactician of Zhao, thinking that the troops of Han Xin must have left provisions behind due to the long march, suggested a surprise troops be dispatched to cut off their supply wagons in the rear. In this way, as long as the troops of Zhao could hold fast to the ramparts, they could surely take Han Xin as prisoner in his desperate fight. Chen Yu, however, thought lowly of Han Xin who had been humiliated when forced to crawl under another people's legs. Given their geographical advantages and greater military strength, he thought they could well defeat Han Xin in an open battle.

Han Xin was overjoyed after hearing about Zhao's arrangement and formulated a singular military strategy: Firstly, he selected a force of 2,000 light cavalry soldiers, each of whom was to carry a red Han flag, proceed along a secret route to the mountains behind the base of Zhao army, conceal themselves, and wait for the proper moment to act: "When the Zhao forces abandoned their fortifications and came in pursuit after our forces, you must enter their walls, tear down the Zhao flags and set up the red flags of Han in their place". In order to increase Zhao forces' underestimation of their enemy, Han Xin sent some troops out to draw up in the ranks with their backs to their river that ran through the gorge, a formation in open violation of the art of warfare. Caught by this trick, Zhao forces on the ramparts roared with laughter. Han Xin initiated a battle voluntarily on the next day and for a long time the two armies fought fiercely. At this point, Han Xin deceptively fled to the forces drawn up along the river. The Zhao forces abandoned their fortifications completely to pursue him. Having no other choice, the army of Han Xin was determined to fight to death and could not be defeated. In the meantime, the surprise force of 2,000 cavalry soldiers rushed into the Zhao fortifications and replaced the Zhao flags with their own. The Zhao forces, unable to achieve a victory, were about to return to their fortifications when they discovered that the walls were lined with the red flags of Han. Concluding that the fortifications had been taken by the Han army, the soldiers found themselves in a dilemma. Filled with alarm, they fled in panic to all directions. At this point, the Han forces closed in from both directions, executed Chen Yu on the banks of Chi River and took the king of Zhao prisoner. This was a great victory over a numerically superior and powerful force. Han Xin acquired an accurate knowledge of his enemy, took advantage of his enemy's underestimation and arrogant attitude, and reversed the situation using ingenious tactics. His victory was bound to earn him resounding fame throughout the country.

Han Xin took Li Zuoche prisoner but treated him with the respect due a teacher. He consulted him with the strategy to attack Yan, to which Li Zuoche suggested that Han Xin should deployed the forces as if he were to attack Yan to intimidate the latter, while sending an envoy to Yan to summon their surrender. Han Xin approved of his plan and set about putting it into action. Without deploying any troops, Yan surrendered.

Defeating Qi

Liu Bang came to Zhao and took command of Han Xin's troops in person. He ordered Zhang Er to man and guard the region of Zhao, made Han Xin the prime minister, and proceeded to attack Qi.

In the third year of Han (204B.C.), on his way to attacking Qi, Han Xin received the word that Li Yiji had been sent to bargain with Qi and that Qi had already submitted. Han Xin wanted to halt his march but was counseled by people against this for two reasons: The king of Han had not issued the royal order to stop the attack of Qi. If Li Yiji had conquered Qi with merely his glib tongue, it would be an accomplishment that you could not match by your victories in the battles. Driven by his selfish motives, Han Xin proceeded with the attack of Qi. The king of Qi, concluding that Li Yiji had betrayed him, had him boiled alive and fled to Gaomi when Han Xi reached Linzi. He turned to Xiang Yu for help, who sent his general Long Ju with a force of 200,000 to aid Qi. The joint forces of Qi and Chu drew up on one side of Wei River, opposite to Han Xin's army. Thinking lowly of Han Xin and eager to distinguish himself, Long Ju was headstrong and impetuous on the battlefield. Han Xin ordered his men to make more than 10,000 bags, fill them with sand and block the flow of Wei River upstream at night. On the next day, he led his army across Wei River to battle and pretended to be defeated by the joint forces. Long Ju chuckled to himself: I always knew Han Xin was a coward and forth with pursued Han Xin across the river. Han Xin then ordered his men to break open the dam of sandbags. The water came rushing down so that the forces of Long Ju were dispersed and Chu army was in a rout. With all speed Han Xin closed in and killed Long Ju, thus conquering and pacifying Qi.

In all these three battles, Han Xin made use of the tactics of surprise attack. In the battle conquering Wei, the effects of surprise attack were achieved by deceiving the enemy. In the battle breaking Zhao, knowing Chen Yu's underestimation of his enemy, Han Xin drew up the forces in the ranks with their backs to their river to lure the enemy out of their fortifications. He then brought the psychological edge from the enemy to his by deceiving the enemy, thus battling against the heavy odds and subjugating Zhao. In the battle defeating Qi, Han Xin exploited the underestimation of his enemy, yet again, by the Qi general Long Ju. He cut off the water flow of the Wei River with sandbags, initiated the attack to lure the army of Long Ju in the pursuit of his army and ordered the sandbags upstream be removed when the Qi troops were halfway across the river. Thus putting the Qi troops in a rout, it became easier for Han Xin to deal solely with the troops of Chu. By killing Long Ju, he sealed this victory. In all these three battles, Han Xin was seen to turn defeat into victories against heavier odds. From the troops of Wei, Zhao and the joint forces of Qi and Chu, the strength of the enemy was ever increasing. Han Xi, however, won in these battles after all. He is thus surely worthy of the term of Fairy Strategist.

APPLICATION OF GIS IN THE TEACHING OF "SHI-JI"

As a teacher of "Shi-Ji" in the university's Department of Chinese Studies, the author will give lectures on Huaiyinhou Biography every year. In the process of reading and teaching of paragraphs about war, the students are found to be incapable either of understanding the use of military strategy and tactics, or of keeping pace with the rapid changes in the names of places. Some maps have been referred to in the process, but the effects are found to be quite limited. In 2013, the author became a member of the interdisciplinary team of the university constituted by Professor Kuo-Hung Huang of the Department of E-learning Design and Management; Associate Professors Pei-lun Lee and Hsin-chun Liu of the General Education Center, applied for a research program of "digital humanities", which, fortunately, was granted. Our hope was to open up new possibilities for the teaching and research of literature and history with the help of GIS and e-learning.

The research program was titled as: "Visualized roam in the space of humanities: Developing Geographic Information System to Achieve Innovations in the Research and Teaching of Literature and History --- GIS-assisted Teaching of "Shi-Ji" - with Powerful Generals as the Example". It is a program centering around the teaching of "Shi-Ji" and aiming to enrich the teaching content *via* GIS. The main reason why chapters about generals were chosen as the subject was that these chapters often involved battles. In "Shi-Ji", the paragraphs about battles were moreover featured by rapid spatial movements and changing place names. For the Grand Historian who had the conditions of the entire China in his breast, he could well have put these down with great ease. For the students, however, they might be thrown into great confusion and become less interested. I used to use maps in my lectures, such as «Atlas series of Shi-Ji» edited and drawn by Chi-fang Huang *et al.*, or «Atlas of Shi-Ji» edited and drawn by the Chinese scholar Xu Panqing. These reference books were proved quite helpful to the teaching. Shame was that some maps were not clear enough or in adequate compliance with the teaching requirements. It was only when the author got access to GIS that he discovered that it was possible to draw the maps concerning "Shi-Ji" by himself.

With the further help of Google Earth, these maps could moreover be three dimensional. Customized in this way, these maps were fitter for the requirements of the course and more helpful to the students' reading and learning. In July 2014, together with another teacher, Pei-lun Lee, I delivered the our preliminary research results in a paper GIS-assisted Teaching of "Shi-Ji" -with Huaiyinhou Biography as the Example on the 20th European Association for Chinese Studies that was held in Portugal. There were scholars from America, Russia and Hong Kong who showed strong interest in the study and hoped that the academic circles could have access to the future results. They also said that they would use this teaching model in their teaching. Although oriented toward teaching, studies of this kind could also benefit relevant studies. It was the wish of the Grand Historian that future readers could be eager to learn, be diligent to think and be understanding of his intentions deep in heart" (Wu Di Benji Zan). In other words, he hoped that readers could pierce through the superficial meaning of words into his intentions between the lines. In the reading of historical books, especially sections about wars, if the readers could grasp the spatial elements, they will have a deeper understanding of the context of the wars, of how the war goes on and the use of military strategy and tactics in the process. It is believed that this will be quite beneficial to both research and teaching.

In the teaching statement, this course of "Shi-Ji" was designed as a major elective course for freshman students. I would read the texts together with the students at the beginning of every session so that they could study "Shi-Ji" directly. When the need should arise, corresponding maps and PPT would be used to help the students to achieve a clearer and deeper understanding of the texts. As an article dedicated to detail the ups and downs in the life of a powerful general, Huaiyinhou Biography was a very long text that could be divided mainly into several sections: Occupying the post of Great General; Conquering Wei, Zhao, Yan and Qi; To divide the country into three kingdoms or not; Being killed as a traitor, *etc*. The lectures were thus also divided according to these sections, featuring text reading and PPT-assisted lectures. When it came to the battles in the conquering of Wei, Zhao, Yan and Qi, it was found that the students would still be puzzled after the text reading and PPT presentation. Later, I found a map on page 66 of «Atlas series of Shi-Ji» edited and drawn by Kuo-liang Hung and Chi-fang Huang (Hung & Huang, 2001).

After integrating this map into the lectures, the students' understanding of the positions of the various States and how the battles went about were clearly improved. However, since this map was not dedicated entirely to Huaiyinhou Biography, it was impossible to mark the specific place names and the locations where the battles took place. In addition, since this map was hand drawn and in black and white, the effects it imposed on the classroom were quite limited. In the «Atlas series of Shi-Ji» edited by the Chinese scholar Xu Panqing, a colorful map was found dedicated solely to Huaiyinhou Biography, which was proved to be more practical in teaching (Xu, 2010). By focusing on the presentation of local conditions, however, this map was found lacking when it came to the entire picture, thus also deficient in teaching. GIS maps, on the other hand, are not only colorful, but also adaptable according to the different perspectives and needs of the lecturer, thus very convenient. The map depicting Han Xin's conquering of the four States, for example, is listed as follows (Fig. **1**):

Maps like this are clear at a glance, thus very helpful to teaching. In addition, if these GIS maps are imported into Google earth, they could be transformed into three-dimensional topographic maps, which could greatly help the teachers when they are explaining the use of military strategy and tactics, as well as the conditions during the battles (Fig. **2**).

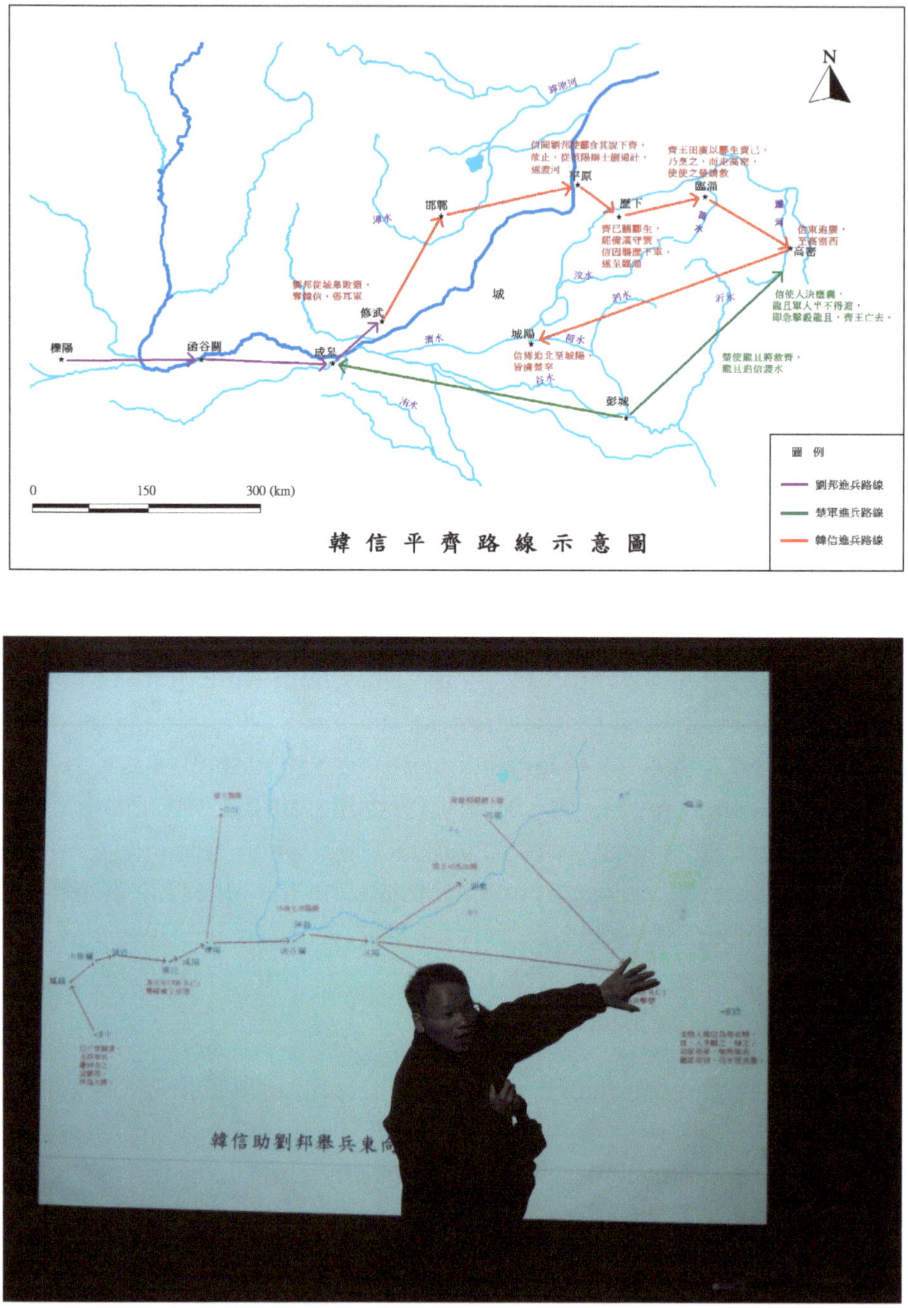

Fig. (1). The layer of GIS and teacher demonstrated in the class.

Fig. (2). Teacher demonstrated with Google Earth and students exercised in the class.

In December 2014, with the help of GIS maps, the author conducted a teaching experiment in the classroom of "Shi-Ji" for the freshman students of the Department of Chinese Literature, Chiayi University: After explaining the paragraphs about war of Huaiyinhou Biography in pure textual forms, a review would be conducted with the help of GIS maps, after which the maps generated by Google Earth on the basis of the GIS maps would also be shown to the students. The homework for the students was also practice-oriented. The students' learning improvements and reactions to this method were collected through pre-tests, post-tests and a questionnaire. 51 students expressed their overall opinions on the course "Shi-Ji" on our open-ended questionnaire.

The questions of the questionnaire given out in January 2015 are as follows:

1. Please give some overall remarks on this semester's course *"Shi-Ji"* in a few sentences.
2. Do you consider the teacher's combination of texts with PPT and maps an appropriate teaching arrangement? Why do you think so?
3. Are the GIS maps helpful to your study? Like stimulating your learning motivation, improving your understanding of the text.
4. In terms of the assignment of this course: namely the adaptation of this biography, do you think of its nature, amount and difficulty as appropriate? Do you find it difficult to complete the assignment on time?
5. Do you have any suggestions for our course?

To the first question "Please give some overall remarks on this semester's course "Shi-Ji" in a few sentences", over 90% of them, that is 46 students, gave the remarks of interesting, ideal. And more than 10% of them thought the course as thought-provoking and causing voluntary reading.

In terms of the "teacher's combination of texts with PPT and maps" of the second question, over 98% of them, namely 50 students deemed it as appropriate. The reason most frequently given out (accounting for 76%) was that this arrangement could facilitate their mastery of the subject in a comprehensive way, thus facilitating their memorization and over 10% of the students considered this teaching arrangement as more vivid due to the extensive supplementary data.

Over 82% of the students, namely 42 people, thought the integration of GIS maps into teaching as mentioned in question three as helpful to their studies in that their understanding of the text was improved, their memory was reinforced, their learning interest was stimulated, *etc*. 17% of them, namely 9 students, however, thought negatively of this arrangement due to their unfamiliarity with GIS and the complex operations, hence the lower learning motivation.

The fourth question was concerned with the overall burden brought by this course, to which 45 students, namely nearly 90% of them considered it just about right.

The fifth question required the students to make some suggestions to the course, to which only 33 students responded. Over 30% of them suggested an increase of teaching hours, while others thought that the interaction between teachers and students, as well as the animation appreciation should be increased. To sum up, this course of "Shi-Ji" had won recognition from the majority of students, and the integration of GIS into teaching had also won the support of over 80% of the students as a means that could facilitate their reading of "Shi-Ji". However, there is still the room for development. For example, the elaboration of GIS and its operations could be more detailed and a method has to be worked out to make sure that the students will operate GIS after class.

CONCLUSION AND SUGGESTIONS

"Shi-Ji" is a monumental work to which the Grand Historian had devoted his lifetime of effort. Given its rich connotations, the number of relevant studies and findings over the past 2000 years is immense. Still, there is much space left to be probed into. The art of warfare was deeply rooted in Sima Qian's profound family education. As the Grand Historian once said, he hoped that future readers could "be eager to learn, be diligent to think and be understanding of his intentions deep in heart" ("Shi-Ji" ·Wu Di Benji · Lun Zan). Readers in the past, have always been limiting themselves to the interpretation of words, without giving due attention to the geographic and spatial elements in the war narratives of "Shi-Ji". They are, therefore, not able to grasp the connotations exactly. Intending the book as a biographic general history, the Grand Historian could well highlight the fascinating lives of the characters. As for the events, they had to be accounted for in the various chapters, to which the careful readers have to refer in order to understand them completely. This paper was thus set out to make up for the fragmented and simplistic manner in which the readers read the paragraphs about war in "Shi-Ji". It was hoped that the readers could understand the information contained in the paragraphs about war in "Shi-Ji" more completely and three-dimensionally. When implemented in the classroom, the method won positive recognition from the students. When delivered as a paper on the international symposium, it had also won substantial support and expectations from the scholars. The GIS-assisted teaching, and research of "Shi-Ji", therefore, is a feasible research direction. This paper is intended only as an initial attempt. In the future, with the help of the time axis and GIS, as well as the interdisciplinary cooperation among the teachers, it is hoped that the paragraphs about war in "Shi-Ji" could be completely organized, so that we could have a deeper understanding of the war narratives, military strategy and tactics, outcomes of war, and the personalities of generals.

CONFLICT OF INTEREST

The authors confirm that the authors have no conflict of interest to declare for this publication.

ACKNOWLEDGEMENTS

This project was supported by the Ministry of Science and Technology (MOST 102-2420-H-415 -007 -MY2 and MOST 102-2420-H-415 -005-MY2).

REFERENCES

Hung, K-L., Huang, C-F. (2001). Atlas series of Shi-Ji. China: Xue-Hai Pub.

Xu, P.C. (2010). Atlas of Shi-Ji. China: Di-Zhen Pub.

CHAPTER 4

Effect of Grid Space Resolution on Historical Data Analysis

Pei-Lun Lee[1] and **Hsin-Chun Liu**[2,*]

[1] *Department of E-Learning Design and Management, National Chiayi University, Chia-Yi, Taiwan, R.O.C.*

[2] *Department of Education, National Chiayi University, Chia-Yi, Taiwan, R.O.C.*

Abstract: This article uses GIS technology to analyze the distribution characteristic of commemorative shrines and steles of the Song Dynasty. The findings show that for the highly discrete historical data, smaller grid is not better. The major distribution is in the southeast coastal administrative regions, and inland Chengdufu Lu and Jinghunan Lu. The hot spot analysis shows that Liangzhe Lu, Huainandong Lu, Jiangnanxi Lu, Guangnandong Lu and Chengdufu Lu have apparent hot zones of construction, and from the north Huainandong Lu to the south Liangzhe Lu, Jiangnandong Lu, Jiangnanxi Lu and Fujian Lu form an apparent belt. The records of commemorative shrines and steles show that Emperor Ningzong of Song has the maximum 75 units, and Emperor Yingzong of Song has the minimum 1 unit. It may be related to the length of reign. In terms of space, there was no obvious location characteristic of distribution in various emperors' reigns in the Northern Song Dynasty, but the major distribution was in the south administrative regions in the Southern Song Dynasty. The results of this study can be provided for history researchers to discuss the historical background and space-time characteristics of the hot zone of construction.

Keywords: GIS, Grid, History, Hot spot analysis, IDW, Shrines, Song Dynasty, Space resolution, Steles, TIN.

* **Corresponding author Hsin-Chun Liu:** Department of Education, National Chiayi University, Chia-Yi, Taiwan, R.O.C.; Tel: 886-5-2717328; Fax: 886-5-2266568; E-mail: hsinchun@mail.ncyu.edu.tw

Kuo Hung Huang (Ed.)

INTRODUCTION

Few history researchers use GIS for history study due to high thresholds of funds and technology, and building a literary history database costs much labor. Therefore, it is difficult to be used for history. However, the space is correlated with time, an additional review of space is helpful to complete analysis of history. For example, Gu (2012) used GIS to establish the distribution of malaria control areas in the Japanese reign, and found that the control areas were not as universal as the epidemic areas of malaria, and the malaria control areas were not all in the high-risk zones of malaria, so the control areas were concentrated where there were dense Japanese, tightly correlated with the nature of colonial medicine. The historical landscapes are reconstructed by using DTM to let the readers know the battle field landscapes or the causes of decisions in the past (Harris, 2002; Knowles, 2004; Rumsey & Williams, 2002) indicated that the GIS actually contributes to clarifying and unfolding historical facts. Therefore, the GIS can integrate the numerous historical materials effectively, the visual data display enables us to analyze and understand regional difference and environmental impact, so as to review historical subjects in a more extensive field.

The historical research on commemorative shrines and steles mostly laid emphasis on the explanation of historical documents, analysis of commemorative stele cases and discussion about administrators' achievements (Liu, 2015; Liu, 2009; Zhao, 2006; Lei, 2004), seldom on space. However, to convert historical data into time dependent spatio-temporal data by combining space with time, and to use GIS spatial analysis to review the spatial layout of historical data will be helpful to understanding the distribution characteristic, so as to analyze the time-space relationship.

Therefore, we use GIS grid analysis technique to research the space dimension of distribution places, hoping to understand the characteristic of spatial distribution, and to analyze the correlation among time, place and neighborhood, so as to look for the hot spot clustering range and recognizable zones in space by hot spot analysis. The data can be provided for literary history researchers to further analyze various location characteristics and the causes of formation, forcing history researchers to research into the time-space relationship of literary history events more comprehensively.

RESEARCH METHOD

The Geographic Information System (GIS) is an integrated system, implementing the creation, access, management, analysis and demonstration of spatial data based on computer. Added to this, it can be integrated with the successively emerging technologies or internet to provide decision making and foreseeing the future by reviewing the past, such as MM, HM, ES, DSS, GPS, RS and Google Earth. The chief values are high flexibility and integration and analysis capability of data processing.

This study uses Mapinfo12.5 as analytical software, there are 442 data of commemorative shrines and steles analyzed, the data are derived from prefectural good governance records in the Song Dynasty (AD960-1278). The time series, personal names, officials, places, administrative content and history sources are displayed in a form. The original construction place is in unit of prefecture. Therefore, this paper uses the prefecture location as locating point, and uses the "Atlas of Chinese History" edited by Tian (1980) as base map to make relevant places and administrative divisions of the Song Dynasty, and corrects the places referring to the layers of Version 2.0, CHGIS of University of Cambridge and Fudan University (Berman, 2004).

In GIS cartographic analysis, the most universal maps are digitized and filed in vector format, such as roads, places and regions. Comparatively speaking, grid data are seldom seen, but they are likely to be calculated and analyzed, so they are mostly used in spatial analysis. Therefore, this study uses vector and grid data formats for related analysis and discussion.

RESULTS AND DISCUSSION

Site Analysis

According to (Figs. **1-3**), the commemorative shrines and steles of the Song Dynasty spread over various administrative regions of the Song Dynasty, but mainly in the southeast. Table **1** shows the top ten administrative regions (Lus) of the Song Dynasty, the top three are Liangzhe Lu, Fujian Lu and Jiangnandong Lu (see Fig. (**1**) for distribution). Fig. (**2**) shows the main construction type is shrine on the base map of current provincial boundary. Fig. (**3**) shows the quantity level distribution in various administrative regions, besides the southeast coastal Lus, the inland Chengdufu Lu is the key region of construction.

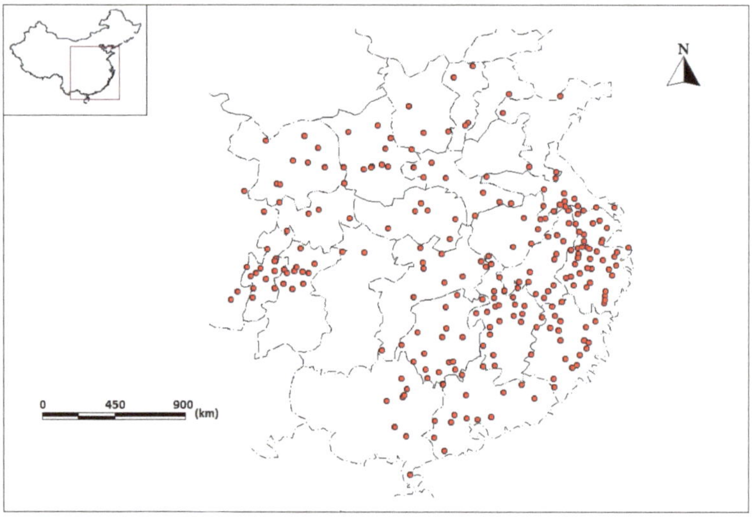

Fig. (1). Spatial distribution of commemorative shrines and steles of the Song Dynasty - taking "Lu" of the Song Dynasty (see Fig. (**3**) for Lu names) as administrative division.

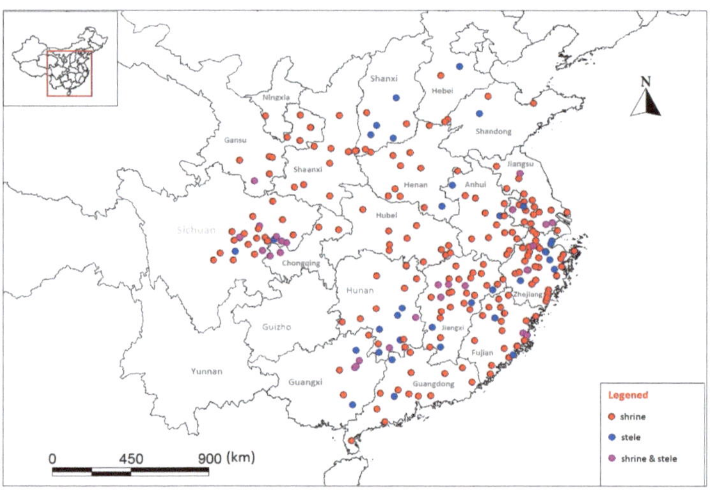

Fig. (2). Spatial distribution of commemorative shrines and steles of the Song Dynasty - taking current provinces as administrative divisions.

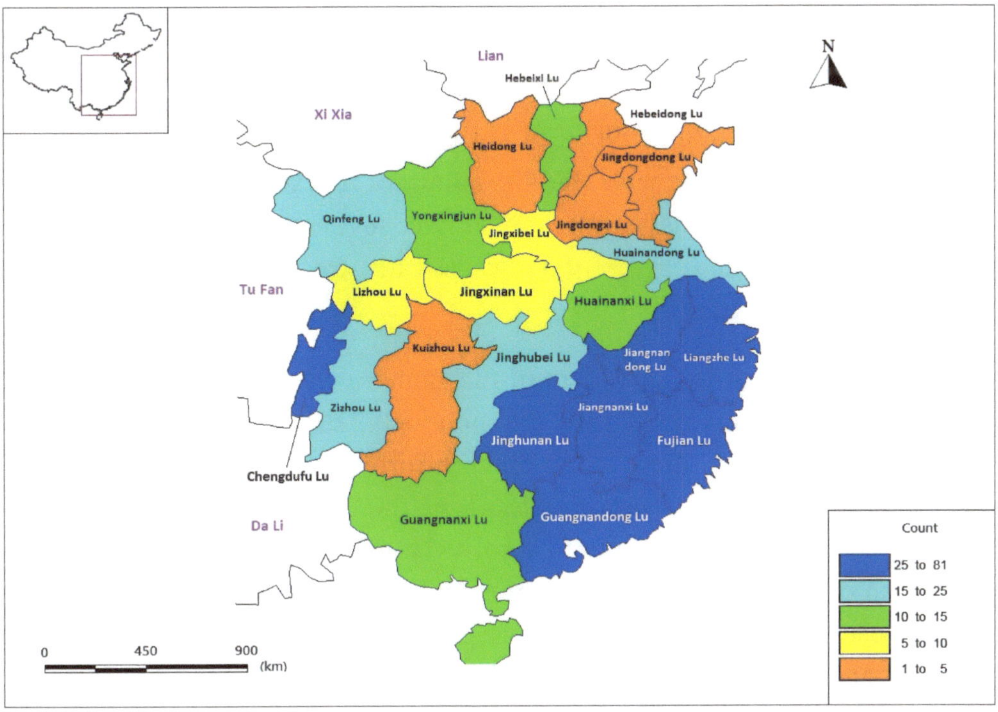

Fig. (3). Quantity range distribution diagram of commemorative shrines and steles in various administrative regions of the Song Dynasty.

Table 1. The top ten administrative regions (Lus) of the Song Dynasty.

Lu name	**Liangzhe Lu**	**Fujian Lu**	**Jiangnanxi Lu**	**Jiangnandong Lu**	**Guangnandong Lu**
Quantity	81	49	40	29	28
Lu name	Jinghunan Lu	Chengdufu Lu	Qinfeng Lu	Zizhou Lu	Jinghubei Lu
Quantity	27	27	21	17	17

Spatial Analysis of Grid Data

The “grid” consists of regular “cells”. The attribute of each “grid cell” is usually represented by the most or the most significant characteristic attributes in the grid cell, but different attributes of the same region can be represented by different grids. Most of GIS analyses or simulations are based on grid-shaped data structure, using grids in a certain size as the basic space unit of analysis. Compared with the vector data which will not deform after scale exaggeration, the grid data are sensitive and disputable in spatial resolution. Openshaw (1984) indicated that the size and combination mode of space unit could result in scale effect and zoning effect, influencing the spatial analysis simulation result. However, Jelinski and Wu (1996) did not regard it as an actual problem in the views of hierarchy and multiscale, but the nature of real world was hierarchical. In other words, smaller grids can record lower-level spatial characteristics of neighborhood or community, large grids only record higher-level spatial characteristics (Huang, 2007).

Generally speaking, for the grid data of the same range, the shorter the grid side length is, the higher is the resolution, the more detailed is the content. However, if the resolution is too high, there is too much information interference, influencing the interpretation. If the resolution is too low, the data are too rough. However, these problems and disputes cannot be solved completely in the present science. Therefore, in order to discuss the relationship between grid size and spatial scale, this paper uses the quantity differences presented by different grid sizes and patterns to analyze the effect on cluster analysis. The research content uses different grid space scales and hot spot analysis to visualize the cluster patterns in different scales and different grid resolutions. However, most of studies use intuition or empirical rule to determine the cell size, so that the analysis result is not clear in explaining the application limit. Therefore, this unit will discuss the effect of cell size and interpolation on historical data analysis.

Nearest Neighbor Distance Analysis

According to the nearest neighbor analysis of the points of commemorative shrines and steles of the Song Dynasty (Fig. **4**), the maximum distance between the two locations is 218.21 km, the shortest distance is 0.09 km, the average distance is about 22 km. It is meaningless to calculate the density of too large or too small grids, the grid is selected based on greater than mean and half of maximum distance, the intervals of 1°, 0.5° and 0.25° latitude and longitude are recommended for making grids.

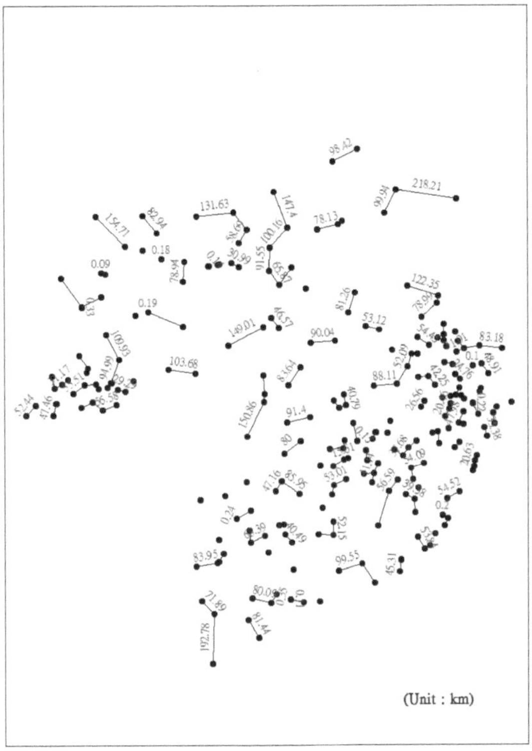

Fig. (4). Two-point nearest neighbor analysis of shrine and stele data (red figures are distance, unit: km).

Effect of Different Cell Sizes on Data Analysis

Due to the result of nearest neighbor analysis, this paper makes grid at intervals of 1, 0.5 and 0.25 latitude and longitude respectively in the original grid division, and uses the contain analysis of SQL directive to calculate the number of points in each grid, as shown in Fig. (**5**). In hot spot analysis, the interpolation calculation is based on 5*5, 12*12 and 25*25km^2 cell size respectively, the interpolation uses TIN and IDW for calculation.

Fig. (**6**) is the comparison diagram of hot spot distributions obtained by analyzing different grid cell sizes and interpolation in the grid in latitude and longitude 1°. As the original grid is large, the spectrum obtained by IDW interpolation inclines to compartmentation, the cluster effect cannot be analyzed. However, if the TIN interpolation is used for hot spot analysis, the 5*5, 12*12 and 25*25km^2 grids present similar spectra, meaning the effect of grid cell size on hot spot analysis is slight. It may because the original grid is much larger than the grid cell.

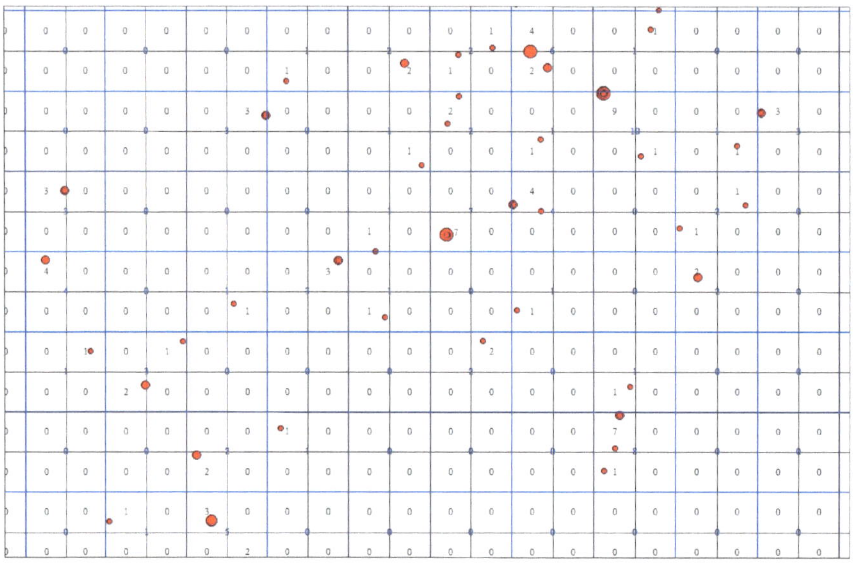

Fig. (5). Number distribution of commemorative shrines and steles in different grid cells.
(Note: blue grid is 0.5°x0.5°, blue figure is the number of commemorative shrines and steles in the grid; black grid is 0.25°x0.25°, black figure is the number of commemorative shrines and steles in the grid. Red circle is the level circle of the number of commemorative shrines and steles).

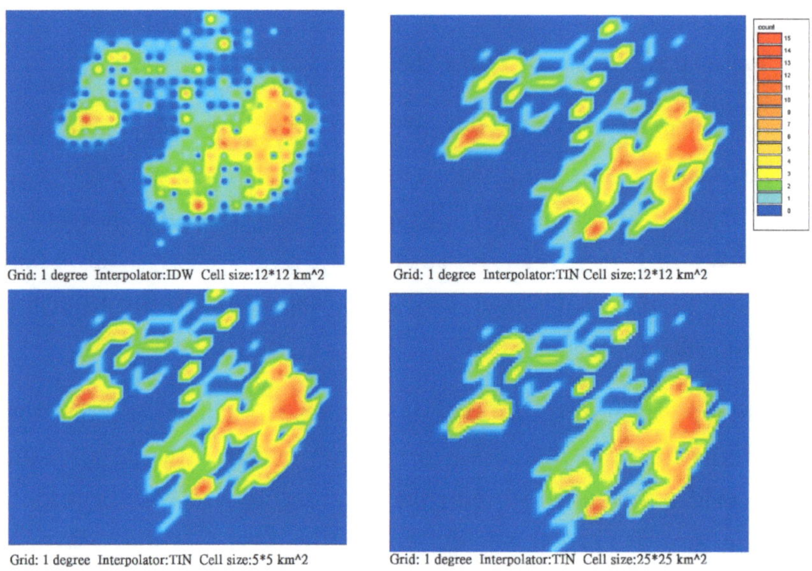

Fig. (6). Hot spot distribution diagram obtained by different cell sizes and interpolation in 1° original grid.

Fig. (**7**) is the comparison diagram of hot spot distributions obtained by analyzing different grid cell sizes and interpolation in the grid in latitude and longitude 0.5°. The spectrum obtained by IDW interpolation is still compartmentalized in the 12*12 km^2 grid analysis, but the distribution position and change of hot spots can be obtained by 25*25 km^2 grid cell analysis. However, relevant hot spots cannot be obtained by using TIN interpolation at this level, mostly independent spots.

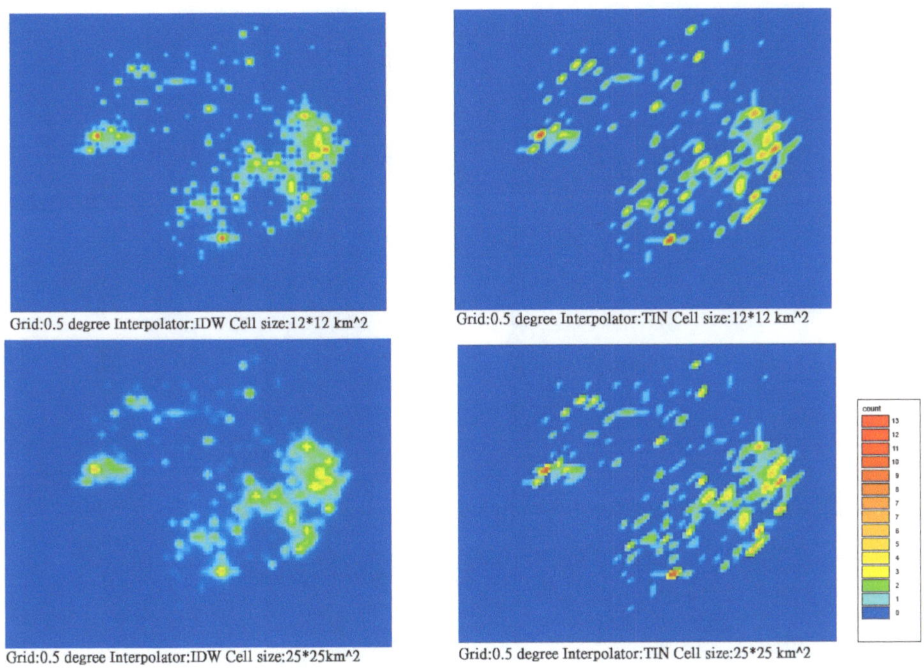

Fig. (7). Hot spot distribution diagram obtained by different cell sizes and interpolation in 0.5° original grid.

Fig. (**8**) is the comparison diagram of hot spot distributions obtained by analyzing different grid cell sizes and interpolation in the grid in latitude and longitude 0.25°. In this original grid, the hot spot distribution cannot be obtained by interpolation or grid cell size, and completely spotted state, meaning smaller grid cell does not represent high resolution.

To sum up, in the hot spot analysis of locations of commemorative shrines and steles, when the original cell size is 1°x1°, the TIN interpolation is more suitable for analysis. Although the 5*5, 12*12 and 25*25 km^2 cell sizes have equivalent hot spot positions in this interpolation, considering the resolution after enlargement, the TIN interpolation and 5*5 km^2 (cell size) are suitable for hot spot analysis in this paper, it is overlapped on spot map, as shown in Fig. (**9**). It is obvious that a lot of regions on the spot map match the hot zones, and Guangzhou has the maximum 13 units among various prefectures.

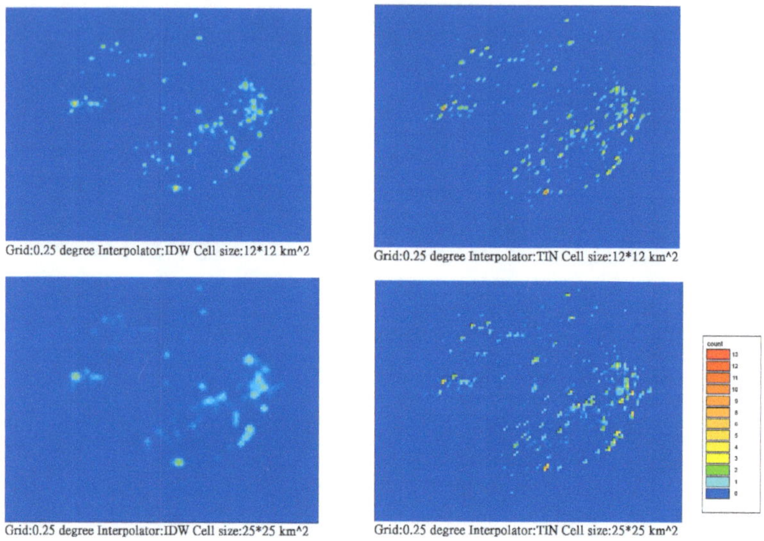

Fig. (8). Hot spot distribution diagram obtained by different cell sizes and interpolation in 0.25° original grid.

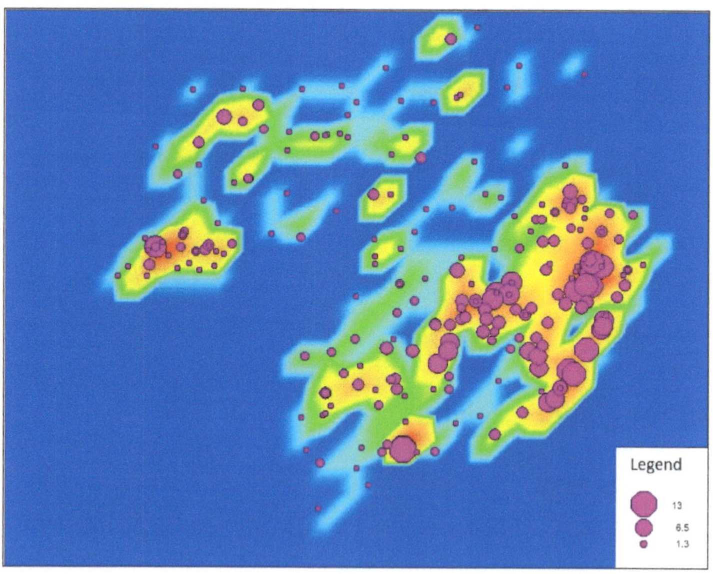

Fig. (9). Overlapped map of spot map of constructions in various prefectures and hot spot distribution.

Hot Spot Analysis

According to the hot spot analysis in Fig. (**10**), the commemorative shrines and steles of the Song Dynasty are located mainly in the southeast coastal area, Liangzhe Lu, Huainandong Lu, Jiangnanxi Lu, Guangnandong Lu and Chengdufu Lu have apparent hot zones of construction. The north Huainandong Lu to the south Liangzhe Lu, Jiangnandong Lu, Jiangnanxi Lu and Fujian Lu form an apparent zonal distribution. The relatively inland Chengdufu Lu has special spatial distribution, the construction of commemorative shrines and steles is very popular in this region. These prefectures are Yan Zhou, Lin'an Fu, Shaoxing Fu, Zhen jiang Fu, Zhen Zhou, Rui Zhou, Fu'an Fu *etc*. According to the above analysis, these zones shall have historical space-time characteristics. This result can be provided for history researchers to discuss the historical background of the region.

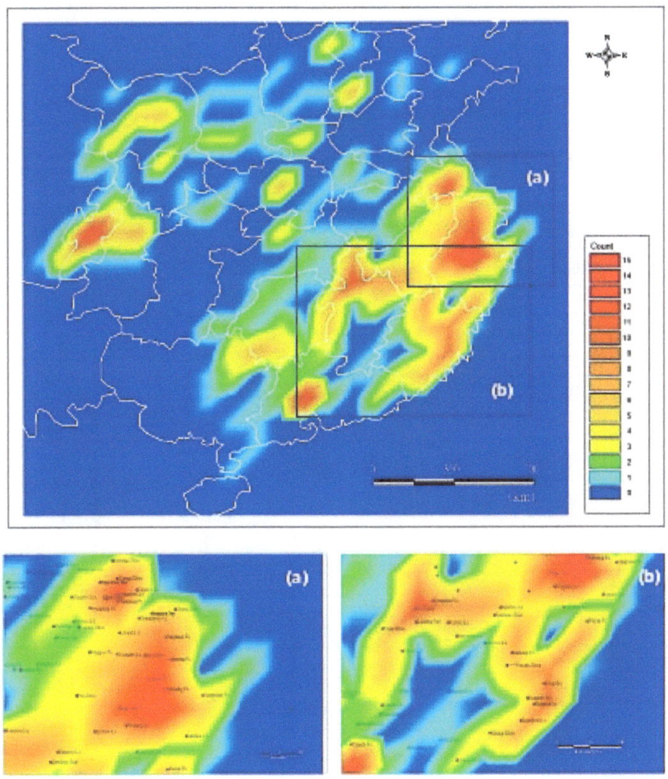

Fig. (10). Red blocks displayed in hot spot analysis are hot zones of construction sites (white line is Lu boundary, Lu names shown in Fig. (**3**).

Analysis of Space-Time Characteristics

There were 14 of 17 emperors of the Song Dynasty who had commemorative shrines and steles in their reigns, but there was no record about Emperor Qinzong of Song, Emperor Gong of Song and Emperor Duanzong of Song. It may because Emperor Qinzong of Song was the last emperor of the Northern Song Dynasty, and Emperor Gong of Song and Emperor Duanzong of Song were the last two emperors of the Southern Song Dynasty when there were more convulsions.

According to the record of commemorative shrines and steles, Emperor Ningzong of Song had the maximum 75 units, Emperor Yingzong of Song had the minimum 1 unit. It may be related to the length of reign. However, the average annual number in the reigns is mostly 1-2.6 units (Table **2**). The spatial distribution is shown in Fig. (**11**). It is observed that the construction sites don't have obvious location characteristic in various emperors' reigns in the Northern Song Dynasty, but they are mostly in the south administrative regions in the Southern Song Dynasty.

Table 2. Record form of commemorative shrines and steles in successive emperors' reigns.

Title of Emperor	Number of Commemorative Shrines and Steles	Years of Reign	Average (units/year)
Emperor Taizu	5	16	0.3
Emperor Taizong	5	21	0.2
Emperor Zhenzong	10	25	0.4
Emperor Renzong	68	41	1.7
Emperor Yingzong	1	4	0.3
Emperor Shenzong	40	18	2.2
Emperor Zhezong	19	15	1.3
Emperor Huizong	28	25	1.1
Emperor Gaozong	62	35	1.8
Emperor Xiaozong	69	27	2.6
Emperor Guangzong	9	5	1.8
Emperor Ningzong	75	30	2.5
Emperor Lizong	48	40	1.2
Emperor Duzong	3	10	0.3

Emperor Taizu	Emperor Taizong	Emperor Zhenzong
Emperor Renzong	Emperor Yingzong	Emperor Shenzong
Emperor Zhezong	Emperor Huizong	Emperor Gaozong
Emperor Xiaozong	Emperor Guangzong	Emperor Ningzong
Emperor Lizong	Emperor Duzong	

Fig. (11). Distribution diagram of commemorative shrines and steles for various emperors in the Song Dynasty.

CONCLUSION AND SUGGESTIONS

1. The results of this study show that smaller grid does not represent better grid resolution. If the grid is too small, the grid chart is net shaped, inapplicable to the analysis. The original grid larger than the mean distance of nearest neighbor by more than two times is applicable to analyzing highly discrete historical point position data.
2. It is observed that the good governance of administrators of the Song Dynasty is distributed in various administrative regions of the time, and concentrated in some regions. The hot spot analysis shows Liangzhe Lu, Huainandong Lu, Jiangnanxi Lu, Guangnandong Lu and Chengdufu Lu of administrative regions in the Song Dynasty have apparent hot zones of construction, and the north Huainandong Lu to the south Liangzhe Lu, Jiangnandong Lu, Jiangnanxi Lu and Fujian Lu form an apparent zonal distribution, the inland Chengdufu Lu has apparent hot zone. In terms of space, the construction sites have no apparent location characteristic in various emperors' reigns in the Northern Song Dynasty, but there are mostly in the south administrative regions in the Southern Song Dynasty.
3. The preliminary result of space-time characteristics of administrative distribution has been obtained by using GIS to analyze administrators of the Song Dynasty, provided for history researchers to discuss the historical background and space-time characteristics. The causes of construction and be analyzed and the point positions of commemorative shrines and steles for administrators of various dynasties can be created in the future, the historical or regional people's administration changes can be observed.

CONFLICT OF INTEREST

The authors confirm that the authors have no conflict of interest to declare for this publication.

ACKNOWLEDGEMENTS

This project was supported by the Ministry of Science and Technology (MOST 102-2420-H-415-007-MY2). And we would like to thank Lin-Jia Hung, Yu-Syuan Lin, Yu-Chun Wang, Kui-Yuan Chen, Pei-Yu Guo, Yu-Ting Zeng and

Min-Lan Fan for their help on the GIS layers work.

REFERENCES

Berman, M.L. (2004). *Key Issues in Compiling a Digital Gazetteer of China's Historical Religious Sites: CHGIS, Version: 2.0.* Cambridge, MA: Harvard Yenching Institute.

Gu, Y.W. (2012). Historical GIS - a new skill of historians. *Academia Sinica Weekly, 1398*, 7-9.

Harris, T.M. (2002). GIS in archaeology. In: Knowles, A.K., (Ed.), *Past Time, Past Place: GIS for History* (pp. 131-143). Redlands, CA: ESRI Press.

Huang, R.W. *(2007). A Study of GIS Grid Resolution*, Master's Thesis Department of Urban Planning, National Cheng Kung University, Taiwan.

Jelinski, D.E., Wu, J. (1996). The modifiable areal unit problem and implications for landscape ecology. *Landscape Ecology, 11*(3), 129-140.
[http://dx.doi.org/10.1007/BF02447512]

Knowles, A.K. (2004). Visualizing Gettysburg: Problems in historical terrain modeling, In: *SSHA Meeting 29th Annual Meeting of the Social Science History Association*, Chicago, IL.

Lei, W. (2004). Levels and operation of local shrines in the Tang Dynasty-centering on commemorative shrines and town god. *Historical Studies, 2*, 27-41.

Liu, H.C. (2009). Tang Dynasty legislating the medium of communication between the central and the local: Discussing the erection of commemorative shrines and steles. *The Journal of Legal Studies, 15*, 1-58.

Liu, Q.L. (2015). The Prefectures and Counties' Officials Political Achievements Written on Benevolent Monuments in the Tang Dynasty. *Journal of Sichuan Normal University, 42*(4), 155-162.

Openshaw, S. (1984). Ecological fallacies and the analysis of areal census data. *Environment and Planning A, 16*(1), 17-31.
[http://dx.doi.org/10.1068/a160017] [PMID: 12265900]

Rumsey, D., Williams, M. (2002). Historical maps in GIS. In: Knowles, A.K., (Ed.), *Past Time, Past Place: GIS for History* (pp. 1-18). Redlands, CA: ESRI Press.

Tian, Q.X. (1980). *Atlas of Chinese History.* Shanghai: Fudan University.

Zhao, K.S. (2006). Analysis of the phenomenon of temple to a living official in Ming Dynasty. *Seeking Truth, 33*(2), 126-131.

CHAPTER 5

Visualizing Place Connection: Application of GIS in Turmoil History Research

Hsin-Chun Liu[1] and **Pei-Lun Lee**[2,*]

[1] *Department of Education, National Chiayi University, Chia-Yi, Taiwan, R.O.C.*

[2] *Department of E-Learning Design and Management, National Chiayi University, Chia-Yi, Taiwan, R.O.C.*

Abstract: The geographic information system (GIS) database contains spatial data from modern and ancient maps, satellite images, natural environments, and artificial scenery, special event, communications network, natural disasters; it even tells the differences between town and country and conflicting events between populace groups from the Song Dynasty. When researches of similar historical themes are lined up together, a GIS with related spatial data and geo-temporal visualization could be designed and implanted. With the aid of GIS, it is possible to draw maps of the Batu Caves Event and to illustrate the cause, the process of and solutions to it. This thesis intends to analyze the boundary management policy within the Song dynasty territory; and the growing state power, and further gains an insight into the governance of the Dong Militia, the economy of Mountain barbarians and also its cultural values. With a literature of Song anthology, epitaphs, and visuals and maps drawn by GIS, this thesis presents a historic scene from 800 years ago and the turmoil in local development.

Keywords: Batu caves, Dong militia, GIS, Mountain barbarians, Provincial people, Regional armies, Savage boundary, Sidong people, Song Dynasty, Subdued tribe.

INTRODUCTION

Because of the high threshold of funds and technology, many historians hesitates

* **Corresponding author Pei-Lun Lee:** Department of E-Learning Design and Management, National Chiayi University, Chia-Yi, Taiwan, R.O.C; Tel: 886-5-2717226; Fax: 886-5-2062328; E-mail:peilun@mail.ncyu.edu.tw

Kuo Hung Huang (Ed.)

to apply the geographic information system (GIS) to their research, and it becomes even harder when building historical databases always takes massive amount of time. Though there is always this difficulty, space and time are closely related and one cannot afford to overlook either one when researching on history. GIS increases the view of space and thus enables a more comprehensive analysis. Therefore, when we look at space and time together simultaneously, the historical data could be transformed into different timing of spatial and temporal data, and this approach helps users to understand the spatial characteristics and allocation patterns of historical data, facilitating an analysis of the spatial and temporal relationship (Mostern, 2011).

History study is always based on large numbers of original materials and solid literature review. In addition, one has to overcome the difficulty of interpreting classical Chinese writings when researching the history of Song dynasty (960-1279). According to ancient materials, Batu Caves was located in the mountain area in Hunan but geographically closer to Jiangxi. The indigenous peoples there were not Han people and were called Sidong people—meaning they were the citizens living in Mountain barbarians' area. During the Batu Caves Turmoil, Dong Militia didn't always hold their ground firmly: they could be anti-government in some cases but also served as local troops suppressing the rebellion in some other. Using the GIS for map drawing, the thesis delves into the cause and process of the Batu Caves Turmoil and also the post-turmoil measures taken by the government; and moreover, this thesis then analyzes the later policy for events that took place near the boundaries between administrative regions within Song's territory. Looking into the growth of government power, this paper offers an insight into the organization and governance of the Dong Militia, as well as the economy and cultural values in the Mountain barbarians' area.

Although the location of Batu Caves has been hand drawn by scholars, those maps are not completely useful for readers to figure out the administration of the Mountain barbarians' area where non-Han people lived (Li, 1969). Some scholars, like Hiroshini (2002) and An (1995), believe Dong Militia only existed along the Zuo River and You River in Guangxi. However, my research points out that the Dong Militia also showed signs of activity in Hunan and Hubei province. For example, the arrow army in Hubei, and the Village Militia and Stockade Militia, are all various types of Dong Militia (Liu, 2008). This paper details how my argument is supported by Batu Caves Turmoil graphics drawn by GIS.

RESEARCH METHOD

GIS is an integrated system that uses computer as a supporting measure that establishes accesses, manages, analyzes and presents spatial data. Moreover, it can be integrated with any types of new technology, especially with the Internet, to provide decision-making and outcome predication. Its main value lies in high degree of flexibility and capacity in integration for data processing and analysis.

Visualizing is distant reading. It makes distance no longer an obstacle but a form of knowledge. With only a few elements, GIS and visualization could pull the tricks and draw forms or models that delineate the shape, relationship and structure that display the entire spatial interconnectivity. Thus, it helps explain research questions more clearly (Franco, 2005).

The theory that Batu Caves Turmoil started as a result of the Luo Shichuan case in 1208 was widely circulated and accepted, but the truth is the droughts that happened in 1206 in various places of Jiangxi, Hunan and Hubei already triggered an unrest and followed by plunders in prefectures and counties. Some might jump to conclusion and simply attributed the cause of Batu Caves Turmoil to either prefectures and counties or the case of Luo, but I am not eager to believe such convenient explanations. Some argue the cause to be Luo Shichuan being a Yao (non-Han people), but it is less likely to be so since he may already pay allegiance to the Song government and become a registered permanent resident. In Luo Shichuan's case where he fought against his uncle, Luo Shi, the verdict that found Shichuan guilty was given by a Han official who held a strong Han ethical idea that emphasizes the younger shall always respect the elder. Later after the turmoil broke out, the tough terrain in Mountain barbarians' areas hindered military events organized by local officials. Because even at places where provincials and non-Han people paid their allegiance to Song government, once the local government showed any signs of weakness, an inevitable decline of governance and stability was surely to come. All in all, we use GIS to analyze a large number of ancient texts and draw maps by using over 120 books and 1113 records. Is it natural disaster? Misgovernment? Or a combining of both?—the essential reason behind the turmoil is to be revealed in this paper.

RESULTS AND DISCUSSION

Analysis of Turmoil Development with GIS

Analyzing the Space of Administrative Region in Chenzhou, Guiyang

When using GIS for analysis, the common maps have to be digitized and turned into vector format first, such as roads, location, regions and so on. For this paper, we used a large number of historical documents to draw ten maps for Black Cave event; these materials include historical records, legal documents, local records, anecdotal novels, epigraph and inscriptions on stone monuments, and local government officials' petitions to the emperor and epitaphs in anthologies of the Song Dynasty. By doing so, we will be able to present a specific image from a certain period of ancient China.

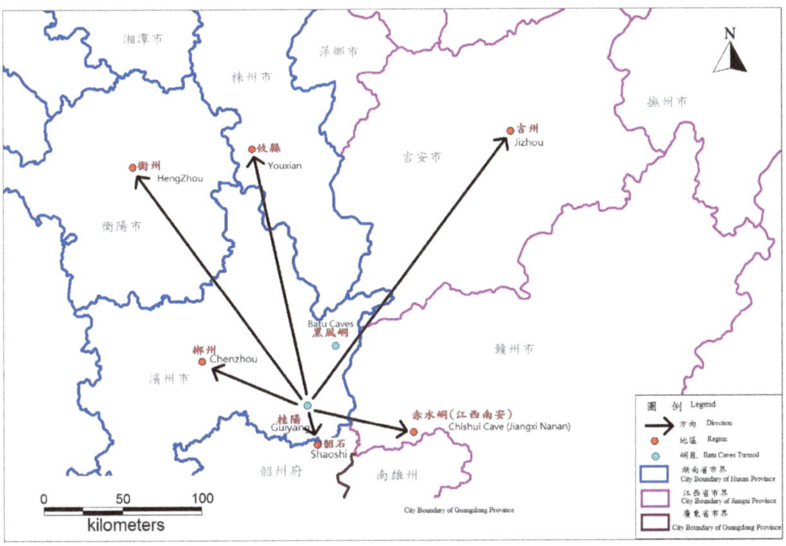

Fig. (1). Source and area map of Turmoil in Mountain Barbarians Region.

Fig. (**1**) was drawn with reference to 18 Collected Works by Song Dynasty

Officials. From the location of Guiyang we could say the turmoil at least spread to six prefectures: eastwards to Jizhou and Nan'an Military Prefecture, Jiangxi province, westwards to Hengzhou, Hunan province, southwards to Shaoguan, Guangdong, and northwards to Youxian, Tanzhou. Fig. (**1**) shows Batu Caves was close to Guiyang, and rebels disturbed up to more than seven administrative regions.

From Fig. (**2**), the topographic map shows that the area around Guiyang and Batu Caves are full of villages and towns. Both Guiyang and Batu Caves functioned almost like autonomous regions due to the difficult natural environment. The Song local officials even once commented, "Batu Caves lies to the east of Guiyang. The mountains between Guiyang and Batu Caves are high and steep. Those are places full of unruly residents and bandits. Because of the tough terrain, residents in Guiyang County of Chenzhou are usually regarded as "stubborn and rude people who tend to fight against the government". Locals, who escaped from their military duty, when unable to sustain their own living, would gather and organize robbery. Fig. (**2**) best presents such reports from local officials in the Song dynasty.

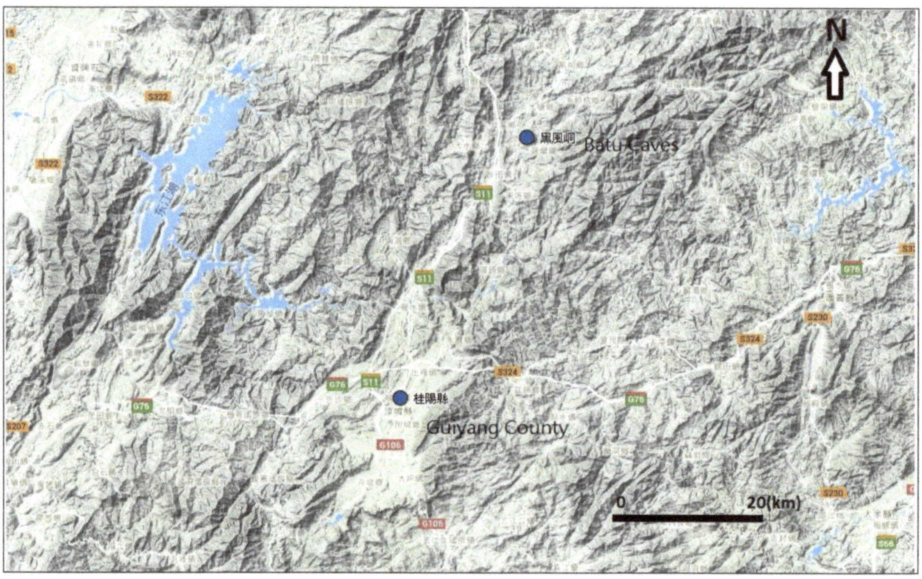

Fig. (2). Topographic map of Guiyang County and Batu Caves (base map: Google map, 2015).

Analyzing the Administration of Hunan and Jiangxi in Song Dynasty

Fig. (**3**) was drawn with reference to 15 local officials' epitaphs and Emperor Ning Zong's biographic sketches. It demonstrates the turmoil was already involved with four prefectures, including Chenzhou, Hengzhou in Hunan, and Ganzhou, Jizhou in Jiangxi. Due to the unsettling and conflicting policies by the officials of the two provinces, Luo Shichuan and his people moved to the Mountain barbarians' area where boundaries were hard to define.

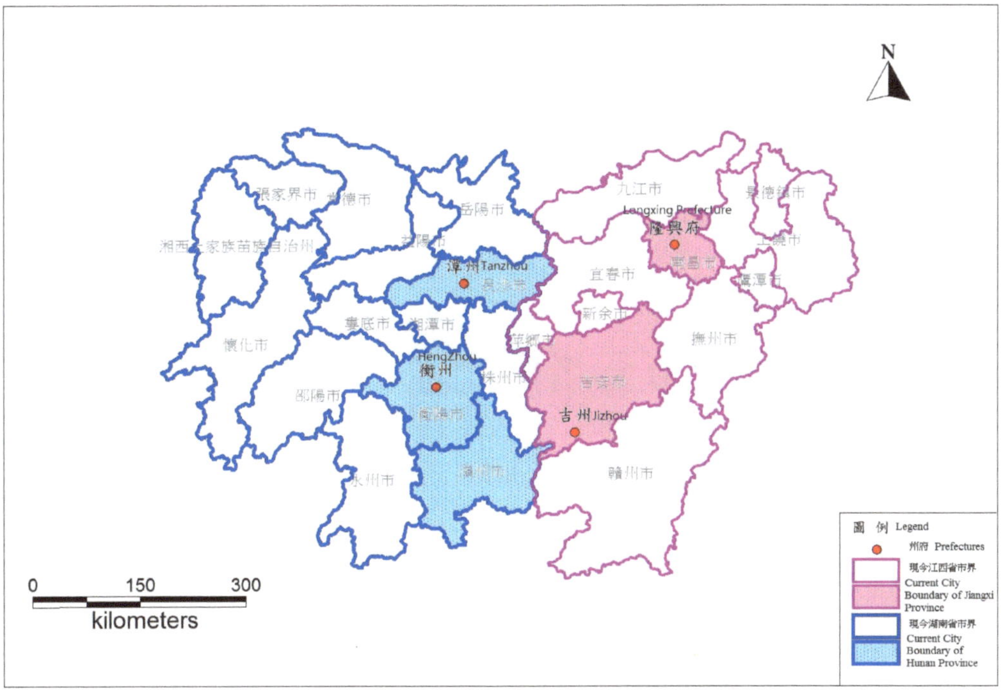

Fig. (3). Administrative map on how government dealt with Dong Militia Turmoil.

Analyzing the Expanding the Area of Batu Caves Turmoil

Fig. (**4**) was drawn with reference to the 17 counter-insurgency wars. Li Yuanli was a Sidong man who took the imperial examination at some point. Since Li Yuanli joined the turmoil and became the leader of Dong Bandits, he made the area affected by turmoil expand northwards and reached Tanzhou, an important administrative region in Hunan province, and also eastwards to Huangshan

determined by a battle fought with the government army. The turmoil also reached southeastwards to Tingzhou of Fujian province as well as Guizhou in Guangxi. Thus, regions brought trouble by the turmoil reached to the number of seven, including the ancient Hubei and Hunan, the east and west of the regions south to the Yangtze River, Fujian, the ancient Guangdong and Guangxi.

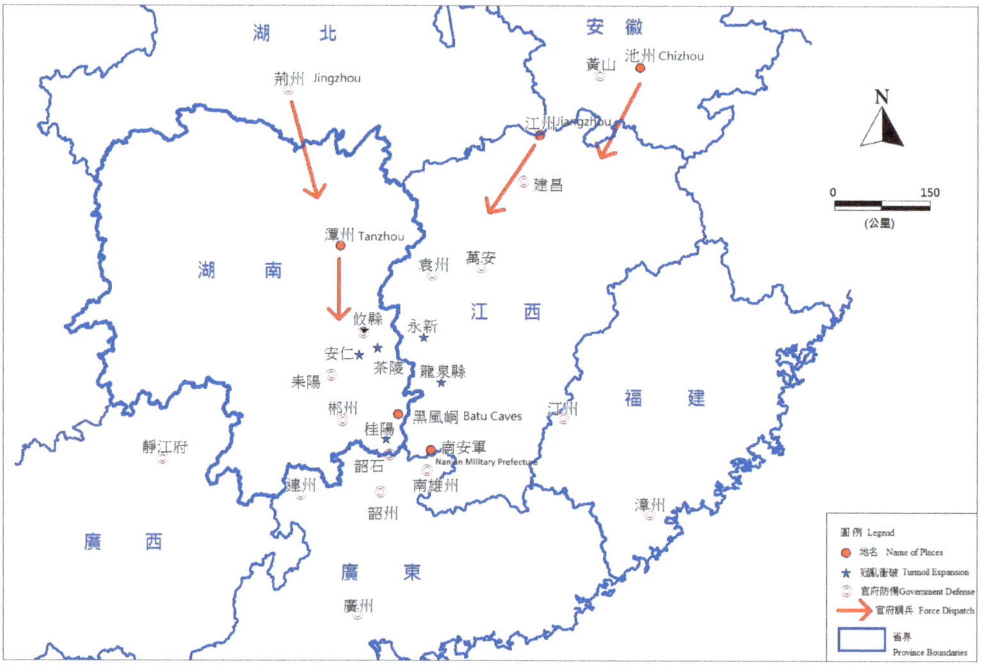

Fig. (4). Expansion of Dong Militia Turmoil and Military Power Distribution.

Analyzing the Consequences of Turmoil with GIS

Analyzing the Increase of Ruled Counties in Southern Song Dynasty

Fig. (**5**) was drawn according to the memorial of the Cao throne. After the Batu Caves Turmoil, the responsible official for the post-turmoil affairs, Cao Yanyue, suggested the government to set up new county governments. Later, new counties like Lin, Guidong and Zixingcounty were established accordingly. Locations of these new governments represent an enhancement of government power in Mountain barbarians' area, helping to solve conflicts among the Sidong people, prevent revengeful murders such as the case of Luo Shichuan, and keep an eye on local bandits.

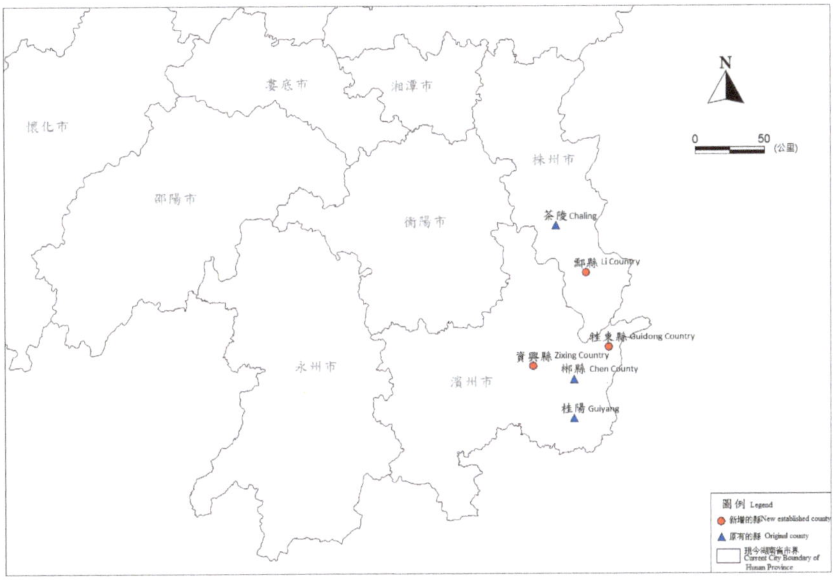

Fig. (5). New administrative map after Dong Militia Turmoil.

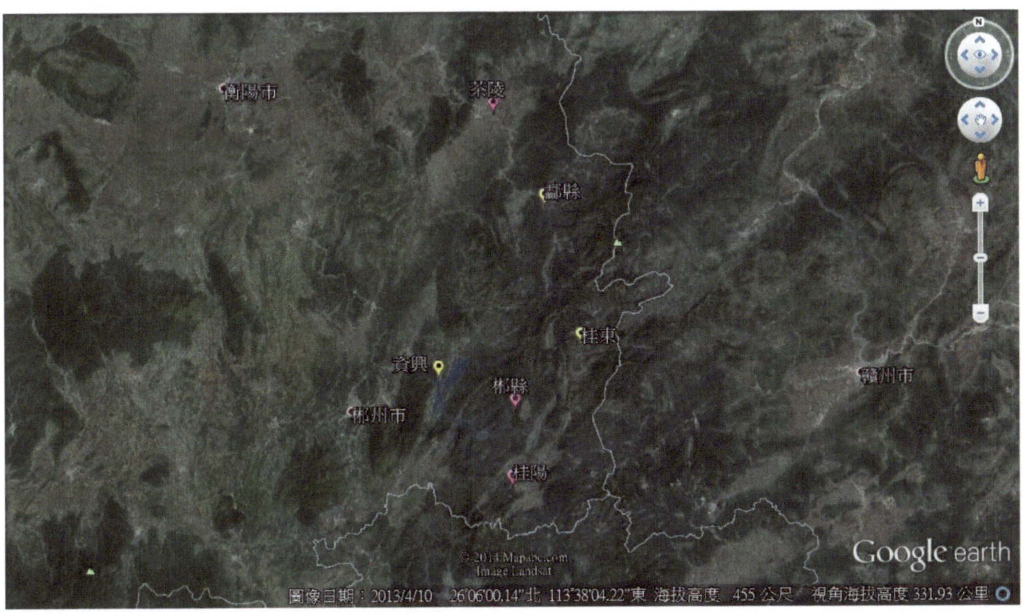

Fig. (6). Administrative and topographic map after Dong Militia Turmoil (base map: Google earth, 2015).

Fig. (**6**) shows clearly the increase of vague boundaries between Mountain barbarians' area and Province, as the latter are subjects to central government taxation. Song government set up administration in villages of Mountain barbarians' area to expand territory that could be directly governed by the central government. The government also had the residence registered and treated the locals as ordinary Provincial people. Since then, strong army was forged with local recruits. With these policies not only did the exchange increase between Sidong people and Provincial people but also trade in farmland, leading to a transformation of the local economy and changing in cultural values.

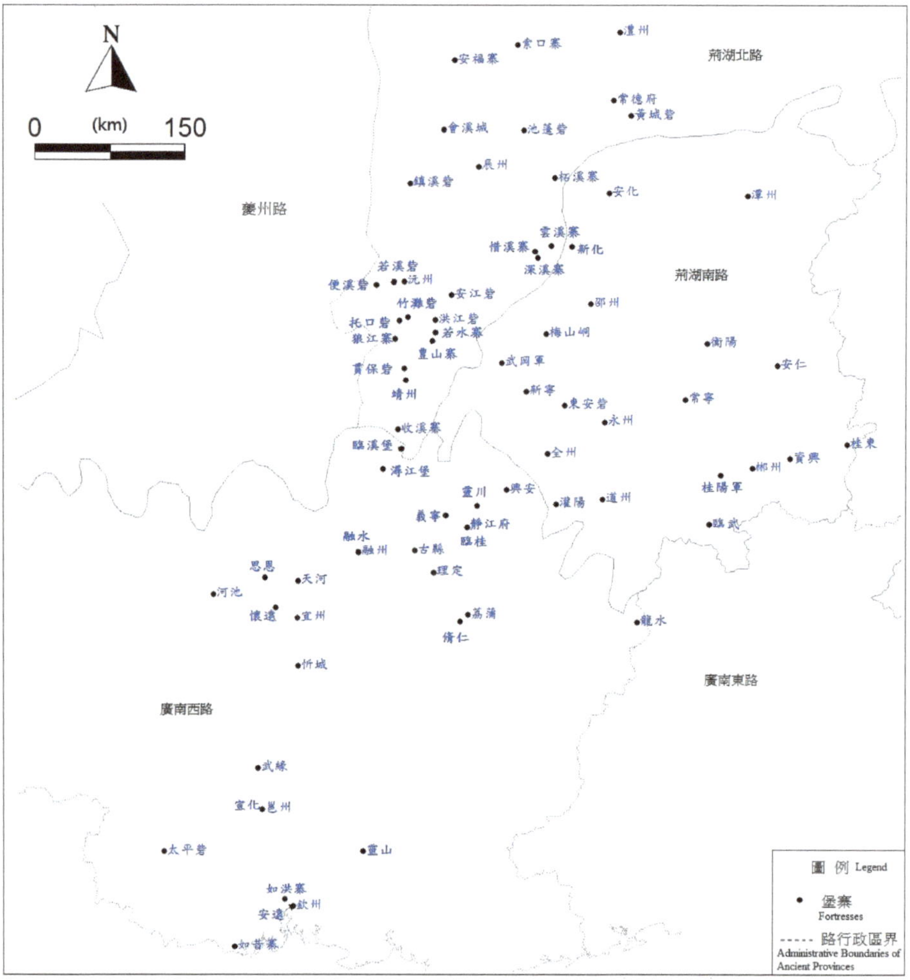

Fig. (7). Prefectures, counties and fortresses where Southern Song Organized Dong Militia.

Analyzing the Song Organized Dong Militia

Figs. (**7-9**) were drawn according to the Monograph on Geography of the Song Dynasty History, including 19 counties and 54 Dong Militias. Figs. (**7, 8**) illustrate the prefectures, counties and fortresses with the administrative divisions of Jinghu North Road, Jinghu South Road and Guangnan West Road as boundaries.

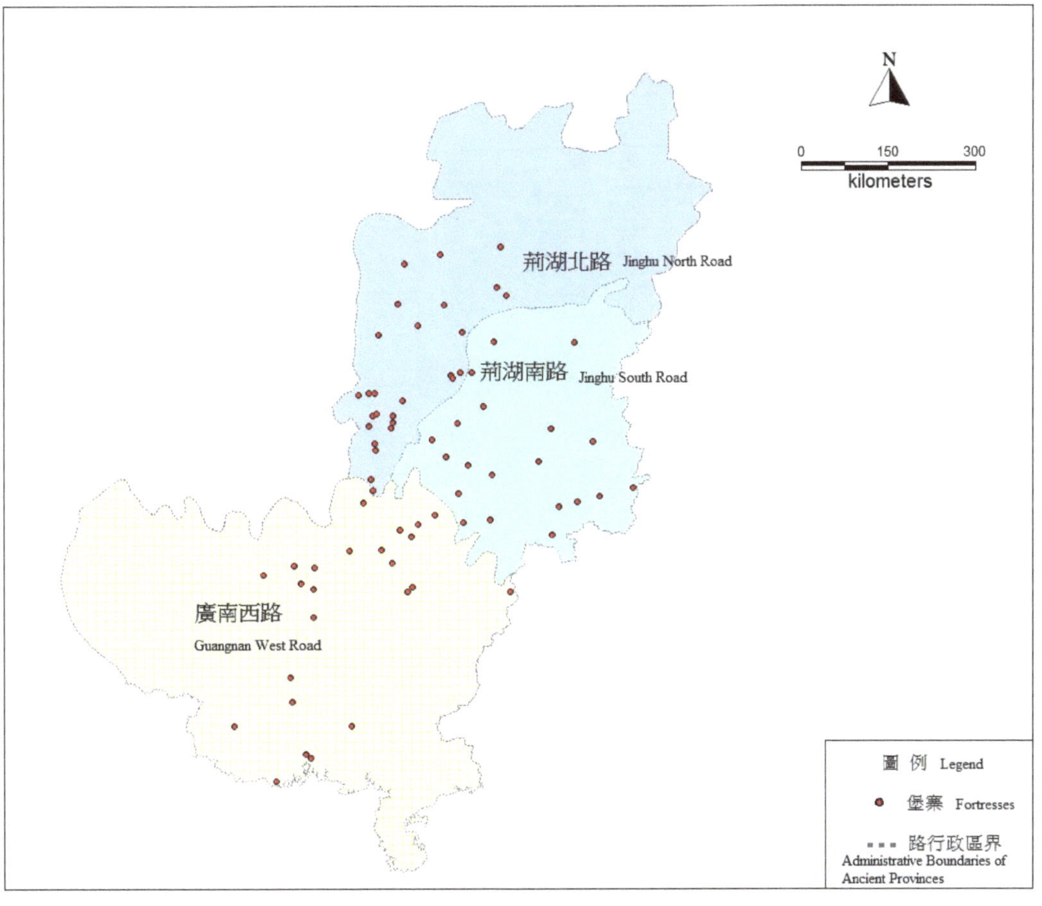

Fig. (8). Prefectures, counties and fortresses where Southern song organized Dong Militia in Ancient Hunan, Hubei and Guangxi.

Fig. (**9**) includes the water areas, autonomous regions, prefectures, counties and fortresses where Southern Song organized events for Dong Militia. The area presented in this figure covers the current Miao Autonomous Region in Xiangxi, Dong Autonomous County in Zhijiang, Miao and Dong Autonomous Counties in Jingzhou, and the Guangxi Zhuang Autonomous Region. Fig. (**9**) also shows the range of Dong Militia was expanded to Ruan River Basin and the surrounding of Dongting Lake, as well as the Zishui River and Xiang River Basin in Hunan and Guangxi.

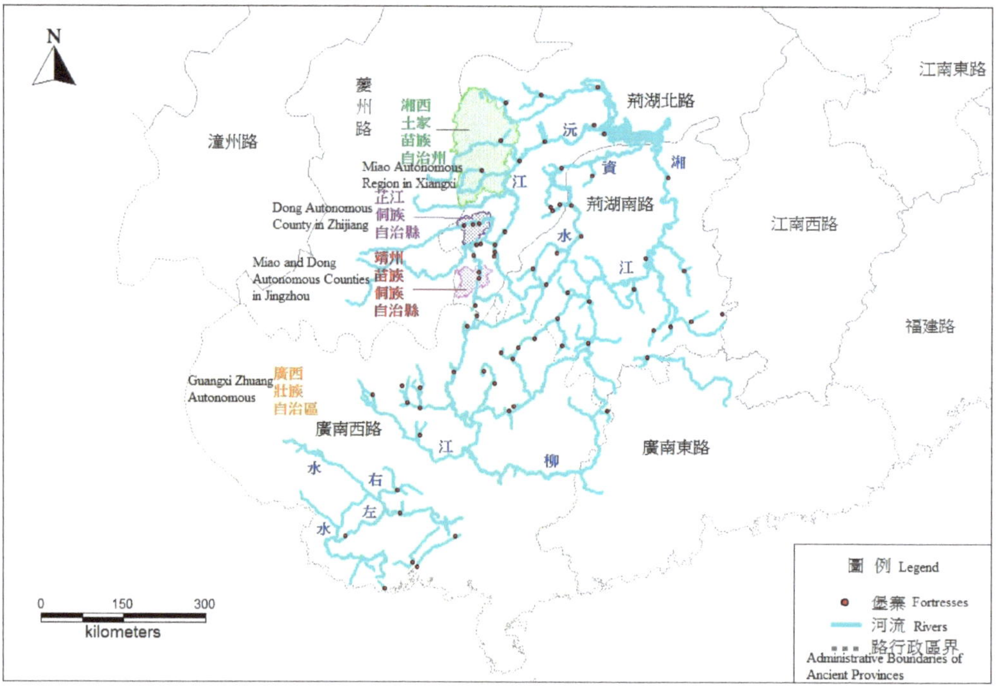

Fig. (9). Water areas and fortresses where southern song organized Dong Militia.

Fig. (**10**) demonstrates the enhanced governance and training of the Dong Militia by the Song dynasty. If one looks at the record in *History of the Song Dynasty*, which states "Dong Militia was seen in Mountain barbarians' area", it could be misleading to think Dong Militia only existed along the two rivers in Guangxi. However, according to regional armies, who were actually registered non-Han residents in Mountain barbarians' area, Dong Militia also refers to the Mountain Tribes in Hunan and Hubei. Most of the Dong soldiers who lived in subdued tribes followed their tribe leaders and took the job as Mountain barbarians'

Officials after paying allegiance to government. The leaders were usually offered posts in government, such as prefectural governor, chief military commissioner, military training commissioner, or magistrates of prefectures, counties and Dong Militia affairs. These are strategies taken to help tackle local problems. Moreover, since the Dong Militia had grown accustomed to the local weather, once they joined the garrison, they could replace soldiers from national army, who were Royal Guards but failed to adapt to local climate. The Dong soldiers who already adapted to the miasma-filled local environment later became an important regional army—a strong force to fight against the Mongolian army for the Southern Song dynasty.

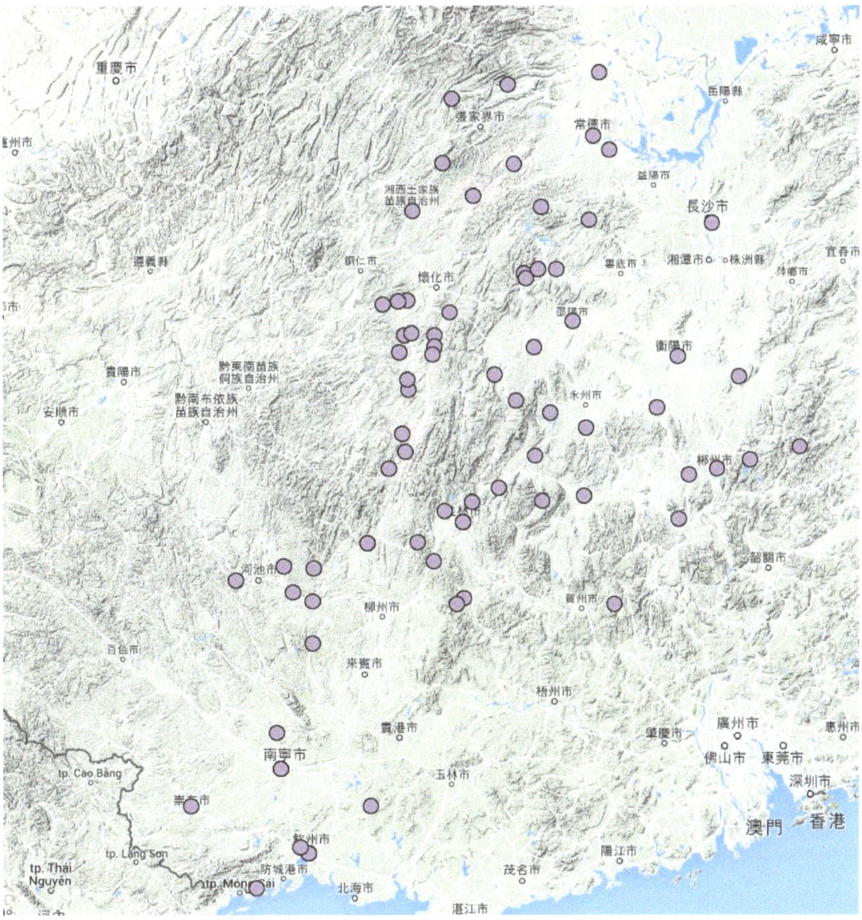

Fig. (10). Topographic map of the prefectures, counties and fortress where Southern Song Organized Dong Militia (base map: Google map, 2015).

CONCLUSION

1. The main goals of this thesis are to visualize history and culture with digital tools and to delve into the events in the past. After sorting out materials from various types of history books, essay collection of Song people and epitaphs, this thesis presents the sprawling and consequences of local turmoil and the struggle between the local government and the Sidong people. With visuals and maps drawn by GIS, it presents a historic scene from 800 years ago.
2. This thesis uses GIS technology to analyze the Batu Caves Turmoil Event. The Batu Caves Turmoil, though it only lasted for merely five or six years in the Southern Song dynasty, had triggered a series of consequences from the early starting point in Mountain barbarians area in Hunan, Jiangxi and Guangdong, to Fujian, Guangxi, and even to places near the plains of Tanzhou. For example, in Jiangxi county governments, there were people pretending to be Li Yuanli, sheltering bandits and lodging false accusations. More crimes of this kind were committed during the quelling of the turmoil where the boundaries between Mountain barbarians areas and provinces became ambiguous, regardless of amnesty policy or the later awarding and punishing measures.
3. With the aid of GIS, this thesis explores the governance of mountain barbarians' area in Song dynasty and analyzes the growth of the government's controlling power, hoping to gain an insight into the organization and governance of the residents in Mountain barbarians' area. The governance and training of the Dong Militia in Hunan and Hubei stood out during the period.

CONFLICT OF INTEREST

The authors confirm that the authors have no conflict of interest to declare for this publication.

ACKNOWLEDGEMENTS

This project was supported by the Ministry of Science and Technology (MOST 102-2420-H-415 -006 -MY2). And thanks Li-Syuan Wu for organizing materials and data.

REFERENCES

An, G.L. (1995). Dong Militia and Volunteer Army in Southwest China in Song Dynasty. *Study of Ethnics in Guangxi, 4*, 26-30.

Franco, M. (2005). *Graphs, Maps, Trees: Abstract Models for a Literary History.* London: Verso.

Hiroshini, O. (2002). *Study of Social History of the Ethnic Groups in South China.* Beijing: Publishing House of Minority Nationalities.

Li, R.T. (1969). The Whole Story of Batu Caves Turmoil- Unrest among Sidong people in Hunan, Guangdong and Jiangxi. *Bulletin of the Institute of History and Philology Academia Sinica, 41*(3), 497-533.

Liu, H.C. (2008). Exercising and Unifying the Dong Militia in Song dynasty. In: Zhu, R.X., Wang, C.Y., Jiang, X.D., Dai, J.G., (Eds.), *Collected Papers on Study of Song History* (pp. 619-637). Shanghai: Shanghai People's Press.

Mostern, R. (2011). *Dividing the realm in order to govern: The spatial organization of the Song state (960-1276 CE).* Cambridge: Harvard University Asia Center.

CHAPTER 6

Developing Non-Playing Network Applications for Japanese Chess

Hiroyuki Tarumi*, **Ryu Miura** and **Toshiki Kinuhata**

Faculty of Engineering, Kagawa University, Takamatsu, Kagawa, Japan

Abstract: Japanese Chess (Shogi) is the most complex board game among chess-like games in the world. Because it is a big challenge, computer scientists, especially AI researchers, have been developing strong programs to play Shogi. However, non-playing applications for shogi have not been studied well.

Human players need various supporting tools for Shogi, including educational applications, research aids, and databases. Especially, post-game discussions (Kansousen, in the Shogi jargon) are very popular among Shogi players, but we do not have well-designed tools for shared discussions on the Internet. In order to satisfy such users' requirements, we have designed a total architecture for Shogi on the net, called SAKURA (Shogi Archives and Kansousen Utilities for Research and Advice). SAKURA has servers for shared database and for discussion management. SAKURA's client software has features to support discussions on game records with shared boards and graphical interfaces to deal with game records with variations of moves. Software interfaces to incorporate AI programs into SAKURA are also defined. Key design issues are database architecture and discussion support features.

SAKURA's environment and tools have been developed as a prototype and evaluated by university Shogi players. In this chapter, we will discuss how Japanese chess can be supported totally by SAKURA. The basic design of SAKURA can be applied to other networked board games.

* **Corresponding author Hiroyuki Tarumi:** Faculty of Engineering, Kagawa University, Takamatsu, Kagawa, Japan; Tel/Fax: +81-87-864-2214; E-mail: tarumi@acm.org

Kuo Hung Huang (Ed.)

Keywords: API, Artificial intelligence, Chess, CSCW, Discussion support systems, Human computer interaction, Learning support systems, Performance evaluation, Shogi, Software platform.

INTRODUCTION

Shogi --- Japanese Chess

Shogi (Russel and Cohn 2012) is one of the chess-variant games, which is very popular in Japan. In Japan, more than 150 professional tournament players are active. Regular TV programs on shogi are weekly broadcasted by NHK (a public broadcasting station in Japan) and a couple of other satellite broadcasting stations. Many cities have private schools of shogi, where people of all ages are playing or learning it. People can also play shogi on the Internet with other human players using game servers (*e.g.*, Shogi Club 24 (n.d.) and 81 Dojo (n.d.)), or play against computer programs on personal computers, tablet PCs, or smart phones.

Shogi is the most complex board game among the chess-variants, because of the following rules. (1) The board size is 9x9, whereas it is 8x8 in case of international chess. (2) Captured pieces can be reused by the opponent player, so that the total number of active pieces does not decrease towards the end of the game. It is a very unique rule that is not found in any other chess-like games. Because of these reasons, Shogi has approximately 10^{226} possible *positions* (*i.e.*, total number of board status). It is much more than the case of the international chess, which is estimated as 10^{120}.

AI Programs to Play Shogi

In case of the international chess, computers won against top-level human players many years ago. The most famous match was the game between "Deep Blue" IBM computer and the champion, Mr. Kasparov, in 1996. However, in case of shogi, computers had not been able to win against human strong players until recent years. Of course the difficulties in developing strong programs are caused by the complexity of shogi.

It has been a good challenging problem for AI researchers in Japan to develop stronger shogi playing programs. After some breakthroughs (given by Bonanza (Hoki 2006), for example), computer programs have been improved. In recent years, AI programs won against top-level professional players many times. In the series of "Den'ousen," games between AIs and human professional players in 2013 and in 2014, computers had more wins than loses against professional players.

It was a big challenge to develop programs to defeat human shogi players. However, because it was too exciting to computer scientists who have interests in shogi, other computer supports for shogi players were left undeveloped.

Requirements

Then, what kinds of supports from computers are required by shogi players? In order to play games, many services on the Internet and game programs are already available. We have focused on other services provided by computers or using the Internet, such as supports for learning shogi and communication between players.

To learn shogi skills, one considerable method is to play game with a strong computer program. However, they are too strong to most of the amateur players. Of course it is possible to tune the strength of AI programs, but even tuned programs do not give instructions to players; they just play badly. It is required to develop interfaces between strong AI programs and average human players, with which players can learn something.

We also focus on *Kansousen*. Kansousen is a Japanese word used commonly by shogi players, which means discussions between the players after playing a game. They discuss the game, examining the moves they played, and other moves they should have played. Typical utterances during the discussion are: "what would happen if I (or you) played this alternate move?" or "which move is better at this position, this move or that one?" A series of supposed moves after an alternate move is called *henka* in Japanese, which means a variation. We call it a *variation* in this chapter. Kansousen is popular among shogi players and it helps players to improve their skills of playing shogi. A good computer-supported environment for Kansousen is required.

Another method to improve shogi skills is to learn game records (*kifu*, in Japanese) given by professional or top-level amateur players, or standard opening moves (*joseki*, in Japanese) from books or magazine articles. It would be helpful if an on-line database covering all game records and standard moves is available. However, information of game records and standard moves are actually given by various resources including books, magazines, almanacs, and on-line web sites that sponsors of professional tournaments often provide. A game record database of professional games exists, but it is exclusively used by professional association members and shogi journalists. Moreover, such databases can deal only with game records as data. Variations are important to learn standard moves, but they are not directly dealt with. Shogi learners need databases covering as much information as possible, dealing with variations as well.

Non-playing Applications for Chess and its Variants

As stated above, most of the research activities related to shogi were involved in developing strong shogi-playing programs. However, in case of the international chess or other variants, we can find several research papers or products of application programs that do not play games by themselves.

Learning Support

Sadikov, Možina, Guid, Krivec, and Bratko (2007) developed a chess tutoring system that make use of data of pondering positions given by AI programs. Guid, Možina, Bohak, Sadikov and Bratko (2013) also gave a chess tutoring system focused on endgames.

Databases

Some voluntary-based or commercial-based databases are available on the Internet. In case of shogi, *KIFU Database* (n.d.) is a popular database for professional game records. In case of the international chess, endgame database and opening database are available (*e.g.*, ChessliB (n.d.) and ChessBase Database (ChessBase Shop (n.d.)).

Application programs using or analyzing databases are also proposed. Wirth and Fürnkranz (2014) tried to analyze annotations on game records to learn players' move preferences. Ganguly, Leveling, and Jones (2014) tried to retrieve "similar" chess positions from an archive of games.

User Interface Technologies

Piškorec, Antulov-Fantulin, Čurić, Dragoljević, Ivanac, and Karlović (2011) developed a computer vision system to recognize the moves of chess game by using two cameras. Wang and Green (2013) used a video stream from one camera for the same goal. Wang and Shi (2014) developed a remote Chinese chess game system using augmented reality techniques.

Lu, Wang, and Lin (2014) developed a system to visualize a game record with variations. It is designed for users to examine the record efficiently, providing global and local views, and giving tactical analysis data.

DESIGN CONCEPT OF SAKURA

Design Policy

We have developed an application platform for shogi players and software on the Internet, which supports game matches, Kansousen, and personal studies of shogi with API (Application Programming Interface) and databases. It is named SAKURA (Shogi Archives and Kansousen Utilities for Research and Advice).

In the design of SAKURA, we took some important design policies. As described in the last section, there are some researches to support players in some "non-playing" aspects, especially in case of the international chess. Our important policies given below make differences from them.

Open Platform

SAKURA should not be a closed system. By disclosing related communication protocols, we let external software make connection to SAKURA's servers.

Extension to Game Observers

Existing shogi systems are usually for players. However, hobbyists are not only playing their games, but also watching other players' games. We will call this type of stakeholders as *observers*, in this chapter. They also try to improve their skills of shogi, by studying standard moves and past game records. It is required that SAKURA can help users to study them.

Position Database

Most of the shogi systems has databases of game records. SAKURA has a database of positions as well as game records. A *position* is a status of game, which is defined by the arrangement of pieces on the board, pieces in hands of both players, and turn of the move (which player moves a piece next).

We have some reasons to introduce the position database. First, without a position database, data of same positions appearing different game records cannot be unified. In case of Fig. (**1**), the two games (ID 150 and 847) have same moves for the opening moves (until the 21st move, for example). The game records are saved in the database with some comments given during their Kansousen. For example, suppose that a comment is recorded at the 19th position of game 150, and another comment is recorded at the same position of game 847. These comments are both useful to help other players to understand the 19th position, but it is not simple to refer to both comments given with both game records, and even difficult to find all comments given with all games that have the same position. By having a position database and aggregate all the comments, it will be helpful to users to refer to all of them.

The second reason is to deal with positions independently from game records. For example, during the study of standard moves or some other problems, hobbyists often consider a particular position, without associating it to particular game records. By providing a position database, the system will help such activities.

Finally, it is helpful to deal with variations in Kansousen. At discussions in Kansousen, variations are proposed by the participants. Positions appearing in variations are not included in the game records in a straightforward manner. A position database will be useful to record variations.

Note that some online databases of shogi provide position data. For example "SHOGI MAZE" (n.d.) gives positions of typical opening games.

Kyokumenpedia (2014) is a much larger database with more than 130,000 game records and 12,000,000 position data. However, they are interactive web services and cannot be able to be used from software.

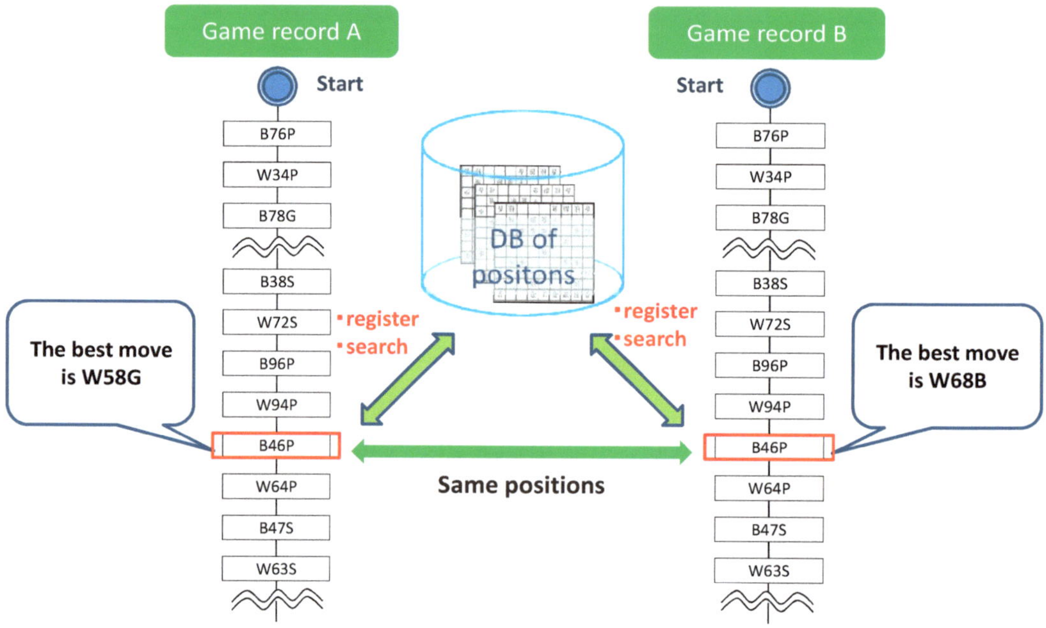

Fig. (1). Same positions appearing in two game records.

Kansousen Features

Kansousen on the Internet is a networked discussion system, which is a typical type of application of CSCW (Computer Supported Cooperative Work). In the design of a Kansousen support system, we discussed some important design issues. Decisions on these issues have been made mainly based on interviews to shogi players in our university, who have experiences of networked shogi games.

Target Users

It is too hard for shogi beginners to do Kansousen, because they should have abilities to reproduce all moves they have made in the game. Hence simply we defined that our target users are middle or highly skilled classes of players who can enjoy Kansousen. In other words, we do not design any supporting functions for shogi beginners, with which they can do Kansousen.

Communication Protocols

In the design of the discussion support systems, we need to define a structure of the discussion and communication protocols to force discussions to follow the structure. In case of Kansousen, discussions should obviously follow tree structures of the moves that appeared on the game, or proposed as its variations. Hence the discussion is constructed with three types of operations: (1) to select a position for discussion, (2) to give a variation branching from a position, and (3) to give evaluations by comments on variations. We have designed three respective protocols to control the discussion.

Shared Information

To design a CSCW tool, it is generally required to design the shared information. In other words, we must define: (1) which information is shared on screen in real time, (2) which information is shared on the database, and (3) which information should be hidden from other participants.

To deal with the first type of information, WYSIWIS (What You See Is What I See) user interface should be designed. In case of SAKURA, the tree structure of variations and a game board is shared with WYSIWIS interface.

The second type of information is stored in the game record database and the position database, with comments given in discussions.

The last type should be carefully defined. According to the interview we had with shogi players, they need a personal game board that is hidden from other participants, on which each player can move pieces to confirm by himself/herself that his/her pondering is correct. They do not want to share variations or comments caused by their trivial mistakes.

Comments should also be hidden from other users, by not saving on the databases, in some cases. It is because usual players are not as skillful as professional or top-level amateur players. They are not confident of their comments and they do not like to open them. Hence we have given options not to save comments on the database.

Communication Media

In case of networked discussions, it should be designed how they can send their comments to other participants. Major options for this design issue are keyboard (chatting) and voice communication. We have selected chatting because we need text data to save comments. However, we have recognized that voice interfaces are also needed by users who are not good at typing. It remains as future work.

External AI Programs

As described earlier, the ability of AI programs to play shogi has been so improved that they can defeat top-level professional players. It is still improving. Our strategy is to provide interfaces to external AI programs, rather than to develop them by ourselves.

In SAKURA, external AI programs have two roles. One is to give advice to human players during Kansousen. Another is to access databases autonomously and append data of comments or new variations onto the database, by pondering position data retrieved from the database.

Design of Advising

According to the interviews with university shogi players, we have found that they do not need full support from AI programs. It is because they want to ponder and discuss by themselves, for the training. Another reason is that AI programs give too much information if not restricted. In our design, AI programs give advices when users request, or when the discussion ends without discussing serious mistakes.

Design of Database Updating

Recently, several professional shogi players adopted moves played by artificial intelligence programs, even in their tournament games. It is evidence that artificial intelligence programs have discovered new standard moves. We are expecting such result by letting AI programs autonomously access our database and give variations. For this purpose, the API is opened to the public for external AI programs to access the database.

DESIGN IN DETAIL

System Architecture

In Fig. (**2**), the total design of SAKURA architecture is illustrated. It consists of Database Server with API, Shogi Server, external AI Engines, and several types of Client Software.

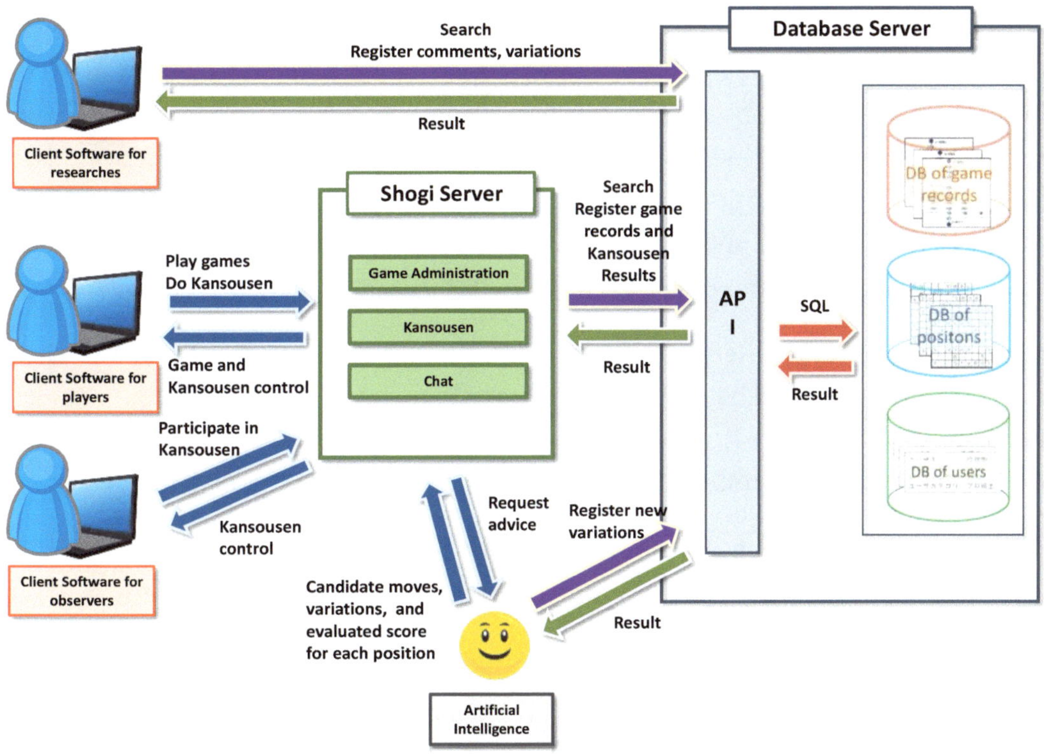

Fig. (2). SAKURA's system architecture.

SAKURA has three types of client software. The first type is the client software for players, with which shogi players can play games and have Kansousen. SAKURA also provides the second kind of client software for observers. Observers on the net will be able to watch games by other players. They can also chat with other observers and are allowed to participate in Kansousen (But currently the second function has not been implemented). The last type of client software is supposed to be used by players personally, in order to study or review shogi games.

Each of the client software is used by players or observers. The clients are connected to the shogi server, which controls the clients during games and Kansousen. The controlling process is defined by our original communication protocols.

Artificial intelligence engines for Shogi, developed by other researchers, have two roles in SAKURA's design: (1) giving advices during Kansousen and (2) proposing new variations and providing comments on positions stored in the database. Currently, we have developed interfaces for Bonanza Feliz (Hoki 2015) artificial intelligence engine. We have also developed interfaces with artificial intelligence programs supporting USI (Universal Shogi Interface) (USI 2015), which can ponder the next move for any given positions.

The databases of SAKURA are comprised of a game record database, a position database, and a user database. The game record database and the position database have links to each other. Records of games, positions, and Kansousen are stored in these databases. The user database keeps records for the user management.

SAKURA provides API to handle databases. All external programs that access the databases must use the API to avoid inconsistent operations, and to keep the program codes simple. At the time of April 2015, more than 220,000 game records (the game records from floodgate (Floodgate 2015)) and positions that appeared in them (more than 27 million) were stored.

Databases and API

Database Structure

Here we focus on the game record and position databases. The game record database collects data of actually played games (game records, player names, winners, venues of games, date and time, *etc.*). It also keeps comments and variations that appeared in each Kansousen. The position database stores data associated with positions, including comments on the position, candidate next moves, *etc*. We have also defined the format of position data (535 bits: 486 bits for the arrangement of pieces on the game board, 48 bits for pieces in hands of both players, and one bit for the turn of game) for the primary key of the position database.

Since these databases have links to each other, Kansousen participants can find comments and candidate next moves on each position appearing in other game records. Consequently, problems explained with Fig. (**1**) is avoided.

The position database has links from each position record to the game records that has the same position, in the game records database. The link data consists of a game ID, a variation ID, and the position number. Consequently, users are able to find the game records that have the same position.

In case of the game record database, it has all position data of 535 bits, each of which is a primary key of the position database by itself.

Variations and Candidate Moves

Variations are stored in the variation table, which is a part of the game record database. A variation is indexed with a game record ID and a variation ID. The two IDs are primary keys for the table. When a variation is stored into the variation table, a variation ID is given with a reference to the position where the variation is branched from the actual game record or from another variation. Variation ID 0 has a special meaning, reserved for the actually played moves. Fig. (**3**) illustrates an example of game record and its variations in discussion.

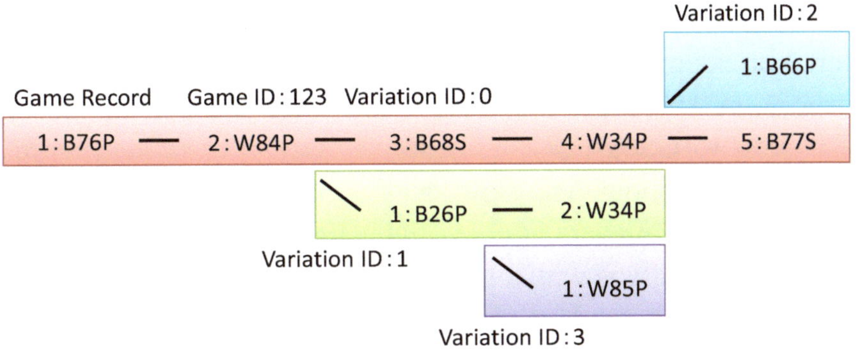

Fig. (3). Example of game record and its variations.

A candidate move means a recommended move or a move likely to be played at each position. Actually played moves on game records and moves given by AI programs with the best score are also regarded as candidate moves. By reference to the position database, users are able to find skillful moves played by other players or standard moves.

Comments

SAKURA keeps two types of comments. They are comments on positions and on game records. In many cases, Shogi learners need comments on positions. For instance, they need to read comments on a particular position when they have questions like: “which player has advantage at this position, and why?”, or “at this position, what are the next candidate moves and which of the moves is the best?”. However, comments that are not useful for learners may also be recorded. For instance, during Kansousen, players may give comments like “I could not find this move of bishop” or “I am proud of this fantastic jump of knight”. Because these comments are subjective, they should not be associated with a position record, but with the game record. The position data should be independent of players. Hence, comments on games including players' subjective sentences should be stored in the game record database. Comments on positions should only have analytical or objective sentences and are stored in the position database. With this design, learners can find only helpful comments on each position when they review games.

Information Related to Users

Data of candidate moves, variations, and comments are to be given with user's information. For example, it includes the data of proposer, his or her rating, and user's category. A “proposer” data item identifies the person who proposed candidate moves, variations, or comments. A “rating” data item represents the player’s skill level of shogi. The skill level is given by integer values, which is commonly used among players of chess-like games. If a player has a better skill, his or her skill level is given with a larger number. A “user category” data item is defined by the authors as one of the six categories: anonymous players, professional players, ladies professional players, apprentices of professional players who are challenging to get professional licenses, amateur players, and artificial intelligence programs.

During Kansousen, players and observers can confirm who proposed the comments, the candidate moves, or the variations, and also confirm the information related to the proposers. Moreover, they can refer to the information of proposers in order to check the quality of candidate moves, variations, or comments. This design to include user categories in the database has been requested by shogi players.

API

It is not easy to operate SAKURA's databases, because of the complex mutual links between databases. SAKURA provides API for the operation from external processes. API functions include the following interface. They are mainly used to access game and position data.

- To save game records.
- To save positions.
- To save variations of a game.
- To save candidate moves.
- To save comments on game records.
- To save comments on positions.
- To update game records.
- To search for game records.
- To search for positions.
- To delete game records.
- To delete variations of the game.
- To delete comments on game records.
- To delete comments on positions.
- To manage user information.

The API is implemented with the PHP language. It communicates with client software, shogi server, AI programs, *etc.*, using the http protocol. API receives POST data from the external processes and sends them XML data.

Client Software

Kansousen Features

Based on the design policies we have already given, the client software for players has two WYSIWIS windows (shared board and shared tree structure), a personal board that is hidden from other participants, and a chatting tool.

Fig. (**4**) is a screenshot of the shared board, used during Kansousen. It has buttons to go forward or back along the game record, to invoke the tree window, to invoke the chat window, to request an advice from AI, *etc*.

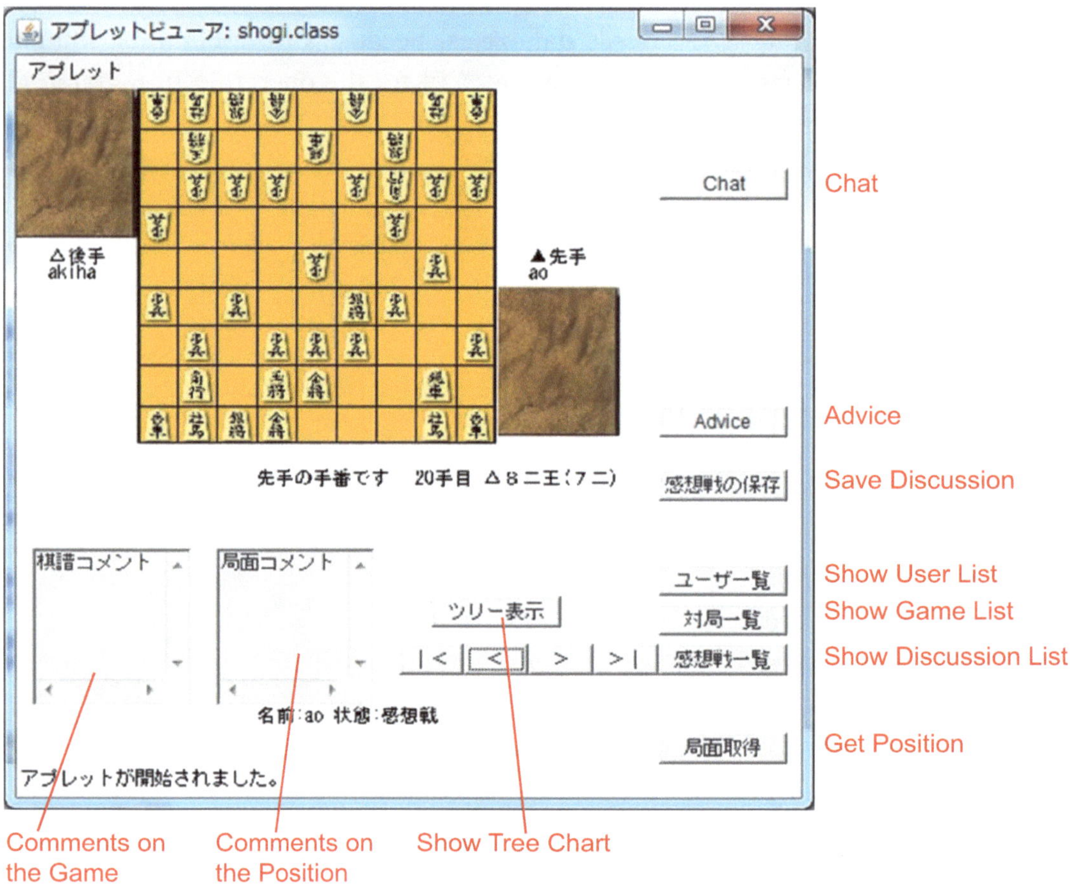

Fig. (4). A shared board interface.

Fig. **(5)** is a screenshot of the tree window. A variation is appended when one participant makes a move on the shared board, which is different from the actual move that appeared on the game. Because it is a WYSIWIS interface, an appended variation is instantly propagated to all participants' tree window.

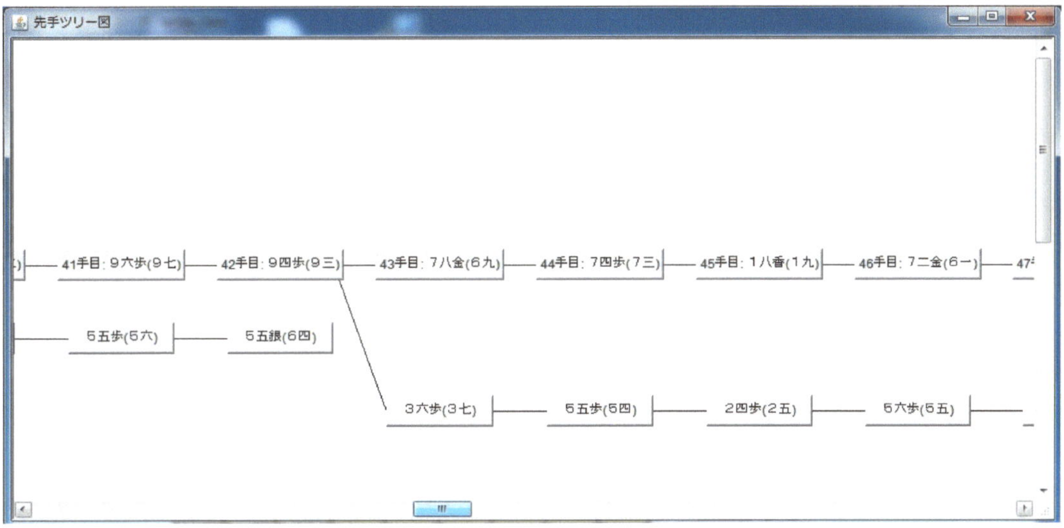

Fig. (5). A shared tree interface.

Comments are given when a participant double-clicks a text on the chat window. The clicked text is extracted as a comment. It is also possible for a participant to type a "#" before typing a text on the chat window, to indicate that the text is a comment attached to the position or the game record. To determine whether a given comment is a comment on the position or a comment on the game record, the participant must select one from the two options, because it is impossible to determine it automatically.

Communication Protocols

As we have already wrote as a design policy, we have defined three protocols between discussion participants to construct the discussion structure. They are to select a position, to give a variation, and to give a comment. These protocols are handled by the shogi server, using the WebSocket and Ruby technologies. WebSocket has been adopted in order to realize push communication to notify the change of status (*e.g.*, a new variation is given) to all related clients.

For example, when one of the Kansousen participants requests a position to discuss, the "select a position" protocol is started. The request reaches the server and the server pushes the request to other participants. If other participants approves the request, the shared game board is changed to show the requested position.

AI Programs

Among several popular AI programs, SAKURA first provided an interface with Bonanza Feliz. With the interface, Bonanza Feliz gives advices during the Kansousen, and also gives new candidate moves at each position. However, the adopted version of Bonanza Feliz (released in May 2010) can only give candidate moves and evaluated scores for an actual game record, but cannot give them for its variations or arbitrary positions. It is simply because it does not support full USI. Then we have developed an interface with artificial intelligence programs supporting USI, with which it will be able to support exchange of arbitrary position data between programs on the Internet.

EVALUATION

Evaluations of Databases and API

Outline of the Evaluation

SAKURA is designed to provide interfaces with AI engines supporting USI. With the interfaces, AI programs can transfer the best move and evaluated score for a given position to SAKURA's position database. During this process, AI programs repeatedly access the database using API a lot of times. It means that the processing-time of API makes significant effects on the total time of database exploration. We made an experiment to figure out the processing-time of the application interface in order to confirm that the interface is efficient enough to support such automatic processing by AI.

In this case, an AI engine executes three steps. (1) It searches positions. (2) It calculates candidate moves and their evaluated scores. (3) It stores them on databases. Hence the process requires three functions of API: the interface to search for positions, the second interface to save candidate moves, and the last one to save comments on positions. The performances of these three API functions have been evaluated.

We have conducted the evaluation experiment on an Ubuntu server, which is a virtual server on Windows 8. The CPU type was Intel R Xeon R CPU E5-2620 v2 @2.10GHz. Two CPU cores and 2 GB memories were assigned to Ubuntu.

The API function to save candidate moves on positions requires arguments of position data and candidate moves. The function to save comments on positions requires arguments of position and comments data. The final function requires only position data as the argument. In the first case, position data and candidate moves are fetched from game records. The position data is also used for other two API functions to stave comments on positions and to search for positions. Just for this testing, we generated random character strings and used them for dummy comments.

The following steps were followed for the evaluation:

1. An empty table of positions was prepared.
2. The loop below was repeated 10,000 times.
    2-1) The API function of saving candidate moves was executed.
    2-2) The API function of saving comments on positions was executed.
    2-3) The API function of searching for positions was executed.

At steps 2-1, 2-2 and 2-3, we measured the elapsed-time of each function, instead of the CPU time. Because the number of data in SAKURA's database was increased by repeating the loop, we checked the increase of elapsed-time associated with the increase of data in the database. We repeated the processes three times.

Results

We measured the elapsed time of each of the three API functions. Fig. (**6**) is the graph of the result of API "to save candidate moves". Because graphs of other two

API functions are quite similar, they are omitted in this manuscript.

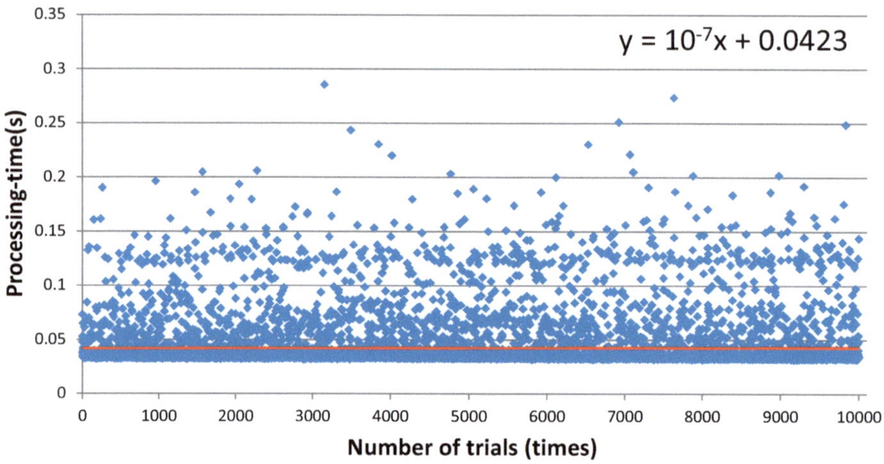

Fig. (6). Elapsed time for "to save candidate moves".

The thick broken line is the regression line, with the regression formulae shown in the graph. The measured data are scattered because they are measured by the elapsed-time, not by the CPU time. The regression coefficients for the number of trials of the formula is small enough and can be neglected. The estimated elapsed-time of the three APIs, to save candidate moves, to save comments on positions, and to search for positions, were 0.0423 sec., 0.0369 sec., and 0.033 sec., respectively. Consequently, the operating time was estimated as 0.1122 sec. for the artificial intelligence programs to execute the process. AI engines usually need more than 10 seconds to compute candidate moves and an evaluated score, but the time depends on the setting of each AI engine. We confirmed that the operating time of the database is enough smaller than AI's computation time.

Evaluations of Kansousen Features

Outline of the Evaluation

Kansousen features of the current version of client software for players have been evaluated by our university's student shogi players. The evaluation session was conducted on January 31th, February 5th, and February 12th, 2015, with the participation of totally seven subjects. They had two to seven years of experience of playing shogi.

An evaluation session was conducted as follows. First, subjects played a shogi game with our client software. After the game, we gave instructions on the Kansousen features of the system to subjects, and let them have Kansousen on the network, without time limit. In these sessions, subjects were in a same room but they were not able to see other participants' display. They were allowed to have oral communication in the sessions, though our basic design is to use text chat. It is because we wanted to focus on the discussion structure and its related interfaces, avoiding the effect of difference of the typing skills. However, they typed when they liked to give comments.

After Kansousen, we asked them to fill a sheet of questionnaire and gave interviews to them.

Results

Here we omit the details of questionnaire and interviews and only give the results.

The responses from subjects were divided into two types. Two of the subjects preferred traditional way of Kansousen and did not like our tool. One subjects said he liked real game boards rather than virtual boards on computers. Another subjects said that he preferred direct face-to-face communication to remote communication. However, their preference cannot be realized when they have to play remote games on the Internet.

Other five subjects preferred SAKURA's Kansousen support features. Especially, the positive evaluation was given at the amount of discussions they could make. With the tree interface and recorded comments, they could have deeper and wider discussions.

The support from the AI program was positively evaluated by all subjects, including the two subjects who did not like this tool.

As for the detailed user interfaces, the subjects were not fully satisfied with SAKURA's tools. One of the reasons was in some remaining faults in

SAKURA's implementation. However, their requirements included visual effects, sound effects, visual design of pieces, time management in playing games, *etc.*, and they were not directly related to Kansousen features. But they requested them because many popular commercial shogi applications provide such features.

Totally, we are considering that the Kansousen features will satisfy shogi players because they can give wide and deep discussions with records and they can receive advices from AI programs. However, we still need to improve the user interface and still need thorough evaluation with more subjects and their longer experiences.

CONCLUSION

Concluding Remarks

We have designed, implemented and evaluated SAKURA, a supporting environment for network shogi games, including non-playing applications like Kansousen support tools and databases. Although we have not yet implemented all functions and tools, the result of evaluation show that the basic design of SAKURA is promising for many shogi players.

Especially, the Kansousen tool is unique and positively evaluated because it helps participants of Kansousen to have wider and deeper discussions. A supporting framework with AI programs was also highly evaluated.

Future Work

The remaining future work includes the following issues.

Tools to Implement

The client software for observers, interfaces to wider range of AI programs, more sophisticated user interface for shogi games and Kansousen, *etc.*, are still left for future implementation.

Continuing Evaluation

Evaluation of the Kansousen tool with more participants, for longer terms, and with more number of games is needed. For the database, we have confirmed that it provides enough efficiency, but we still need continuing evaluation for open use by many users.

Utilization of AI Programs

We have not yet tried any experiments to let AI programs make repeated access to the database and give comments and variations automatically. This experiment will be interesting because we can expect to find new standard moves by artificial intelligence.

CONFLICT OF INTEREST

The authors confirm that the authors have no conflict of interest to declare for this publication.

ACKNOWLEDGEMENTS

We appreciate comments from Ms. Madoka Kitao (lady's professional shogi player) and Mr. Tomohide Kawasaki (81dojo.com). We sincerely thank Prof. Tomoyuki Kaneko, for his cooperation in providing data from floodgate.

This research is partially supported by KAKENHI (25540169).

REFERENCES

81 Dojo (n.d.). *81 Dojo* Retrieved from http://81dojo.com/

ChessBase Shop (n.d.). *ChessBase Shop* Retrieved from http://shop.chessbase.com/

ChessliB (n.d.). *ChessliB* Retrieved from http://folk.uib.no/pfvaf/chesslib/

Floodgate (2015). *Floodgate* (in Japanese). Retrieved from http://wdoor.c.u-tokyo.ac.jp/shogi/floodgate.html

Ganguly, D., Leveling, J., Jones, G.J. (2014). Retrieval of Similar Chess Positions. *Proceedings of Special Interest Group on Information Retrieval, 14*, 687-696. [Gold Coast, Queensland, Australia, ACM.].

Guid, M., Možina, M., Bohak, C., Sadikov, A., Bratko, I. (2013). Building an Intelligent Tutoring System for Chess Endgames. *Proceedings of the 5th International Conference on Computer Supported Education (CSEDU 2013)* (pp. 263-266). Aachen, Germany.

Hoki, K. (2006). Optimal control of minimax search results to learn positional evaluation. *Proceedings of*

Game Programming Workshop 2006, Information Processing Society of Japan.78-83.

Hoki, K. (2015). *Bonanza – The Computer Shogi Program.* Retrieved October 9, 2015 from http://www.geocities.jp/bonanza_shogi/

Kifu Database (n.d.). *Kifu Database* Retrieved from http://kifdatabase.no-ip.org/shogi/

Kyokumenpedia (2014). *Kyokumenpedia.* Retrieved from http://kyokumen.jp/

Lu, W.L., Wang, Y.S., Lin, W.C. (2014). Chess Evolution Visualization. *IEEE Transactions on Visualization and Computer Graphics, 20*(5), 702-713.
[http://dx.doi.org/10.1109/TVCG.2014.2299803] [PMID: 26357293]

Piškorec, M., Antulov-Fantulin, N., Ćurić, J., Dragoljević, O., Ivanac, V., Karlović, L. (2011). Computer Vision System for the Chess Game Reconstruction. *Proceedings of the 34th International Convention on Information and Communication Technology, Electronics and Microelectronics (MIPRO 2011),* Opatija, Croatia870-876.

Russel, J., Cohn, R. (2012). *Shogi..* Berlin, Germany: Book on Demand Ltd.

Sadikov, A., Možina, M., Guid, M., Krivec, J., Bratko, I. (2007). *Automated Chess Tutor.* Springer: Lecture Notes in Computer Science.

Shogi Club 24 (n.d.). Shogi Club 24 Retrieved from http://www.shogidojo.com/

SHOGI MAZE (n.d.). SHOGI MAZE.Retrieved from http://shogimaze.free.fr/

USI (2015). USI Protocol. Retrieved October 9, 2015 from http://www.geocities.jp/ shogidokoro/usi.html

Wang, H., Shi, M. (2014). Recognizing Chessboard and Positioning Checkboard in Chinese Chess Game System *Proceedings of the Fourth International Conference on Communication Systems and Network Technologies (CSNT 2014),* Bhopal, India: IEEE.1182-1186.
[http://dx.doi.org/10.1109/CSNT.2014.240]

Wang, V., Green, R. (2013). Chess Move Tracking Using Overhead RGB Webcam *Proceedings of 28th International Conference of Image and Vision Computing,* Wellington, New Zealand: IEEE.299-304.
[http://dx.doi.org/10.1109/IVCNZ.2013.6727033]

Wirth, C. (2014). Preference learning from annotated game databases. In: Hassani, M., Beecks, C., Seidl, T., (Eds.), *CEUR Workshop Proceedings, Vol. 1226*, 57-68.

CHAPTER 7

An Analysis of the Distributed Leadership Practices in an Elementary School: Focus on the Application of an Official Document Information System

Chin Tu Liang*

Sishih Elementary School, Changhua County, Taiwan; Graduate Institute of Education, National Chung Cheng University, Taiwan, R.O.C.

Abstract: The purpose of this study was to analyze the distributed leadership practices in an elementary school and focus on the application of an official document information system (ODIS). A case study was selected, and participatory observation, interviews, and documents were adopted as the instruments for collecting data. After analyzing and discussing the data, we identified four leadership practices moving toward a professional creative system, where the leaders have positive intervention characteristics, from passive to active followers, a somewhat productive trait on the practices of distributed leadership, and the important finding that the ODIS is a key factor for the practices of distributed leadership.

Keywords: Artifact, Distributed leadership, Elementary school, Follower, Information system, Leaders, Leadership practice, Multiple leaders, Official document, Situation.

INTRODUCTION

Most researches on educational workers and policy decision-makers have studied the issues of school leadership in recent years (Leithwood & Duke, 1999). In such

* **Corresponding author Chin Tu Liang**: Sishih Elementary School, Changhua County, Taiwan; Graduate Institute of Education, National Chung Cheng University, Taiwan, R.O.C.; Tel: +886-19013755; Fax: +886-47786287; Email: lingjindu@gmail.com

Kuo Hung Huang (Ed.)

researches, leadership style usually centers on the leadership mode of a single person, but this approach has been replaced with the perspective that leadership is able to distribute to other staff (Hulpia, Devos, & Rosseel, 2009). In other words, distributed leadership stresses that everyone could lead the organization and its practice transforms formal leaders into informal ones and focuses on the interactive network of leaders, followers, and school situations (Spillane & Diamond, 2007). Spontaneously, it could be inductive to the accomplishment of the complicated projects and tasks and to the improvement of organizational effectiveness.

Recent studies on distributed leadership have focused on the theories of interaction among leaders, followers, and situations (*e.g.*, Spillane, 2006; Spillane, Halverson, & Diamond, 2004), while others have sought to analyze the relationship between teaching accountabilities and teachers' professional development with it (*e.g.*, Harris, 2005; Hulpia *et al.*, 2009; Law, Galton, & Wan, 2010; Lima, 2008). In addition, some studies have centered on the actions of leaders-plus (Spillane, 2005), ignored the need to understand deeply the trust and interpersonal relationships among staff members in school (Gunter, 2003; Lloyd, 2005) or explored the artifacts about the situation (Hutchins, 1995), routines, and tools in school. Thus, they impacted the effectiveness of distributed leadership practice (Spillane, 2006). Relatively, few studies have investigated the interaction among leaders, followers, and situations in day-to-day operational activities (Gronn, 2002) or understand its practice focused on the application on artifacts. In this vein, the current study aims to explore the practices of distributed leadership through the application of an official document information system (ODIS) in an elementary school.

DISTRIBUTED LEADERSHIP AND ODIS

This section examines distributed leadership and ODIS, focusing on the relationship between them, to carry out the subsequent procedures in this study.

Distributed Leadership

Based on the above, *via* distributed leadership, staff can interact with the situation and followers, denoting everyone can be leaders in school, and the school's accountability can also be promoted or changed. First, the practice of distributed leadership is often analyzed through the interaction among the situation, leaders, and followers. However, from the distributed perspective, the situation does not simply "affect" what school leaders do as some sorts of independent variable; it also constitutes leadership and management practice (Spillane & Diamond, 2007). That is, it not only opens the contexts of leadership practice (Spillane, 2006) but also decides the roles between leaders and followers (Lumby, 2013; Timperley, 2005). Moreover, it advances accountability in leading activities, yet it is transformed in the leadership practices (Spillane & Diamond, 2007; Spillane *et al.*, 2004). More specifically, situation has several connotations for distributed leadership, such as artifacts, language (Spillane *et al.*, 2004), routines, and tools in the school (*e.g.*, data on students' test scores) (Spillane, 2006).

In addition, using their available operational power, leaders can achieve the leadership goals (Lumby, 2013). For example, they influence followers' positive actions and other knowledge (Spillane & Diamond, 2007). Either formal leaders or others as informal leaders in the organization can be perceived as leaders in distributed leadership (Hulpia, Devos, & Keer, 2010; Law *et al.*, 2010; MacBeath, Oduro, & Waterhouse, 2004), such as the director of a division or teaching expert in designing curriculum (Lima, 2008; Lumby, 2013).

Finally, the followers are not only an outside variable affecting what leaders do, but also key factors to constituting the leadership practice (Spillane *et al.*, 2004). Furthermore, to reach common organizational goals, they could provide more support to leaders (Timperley, 2005), and their expertise could influence leaders' actions and the situation in the school (BALOĞLU, 2012; Sergiovanni, 2006; Spillane, 2005; Yekovich, 1993).

According to these analyses, the practice of distributed leadership emphasizes the situation, leaders, and followers as critical factors that can all impact accountability in the school's leadership practice.

Official Document Information System

Regarding the form of official paper documents adopted by staff in schools, the ODIS in the operations of the school administration has been accepted due to the technology era. To wit, the latter is able to regard a situation, which has several traits as an artifact. For example, it possesses the social interactive functions for communication and realization with the school task and staff members' other cognitions; furthermore, it interacts with leaders and followers in the operations of the school administration. The ODIS is adopted in this study as a medium or vehicle for understanding what both leaders and followers do and how they use and take steps to analyze the distributed leadership in the school.

Fig. (1). The full-time staff member responsible can understand the processes of documents in ODIS.

ODIS includes several specialties, such as promoting work effectiveness, facilitating administrative operations, and adding news announcements or group meetings (Tsai, 2010). First, in contrast to the ineffectiveness of handling official paper documents by one staff, requiring five to seven days in the past, nowadays only three to five days is required for the same work. In particular, multiple staff members can communicate simultaneously *via* the internet, but staffs still use this work with one another in order to smoothly operate it. For example, the responsible staff member assigns the documents on the internet to one member to complete the work; this individual must handle it and communicate with related to staff members about it. After the work is complete, the system automatically delivers it to the principal for the final decision. More specially, the process of managing documents is supervised by full-time staff (*i.e.*, a key individual designated by the principal) (Chau, 2005). In sum, the process of handling the document in ODIS is very quick and effective (Figs. **1** and **2**).

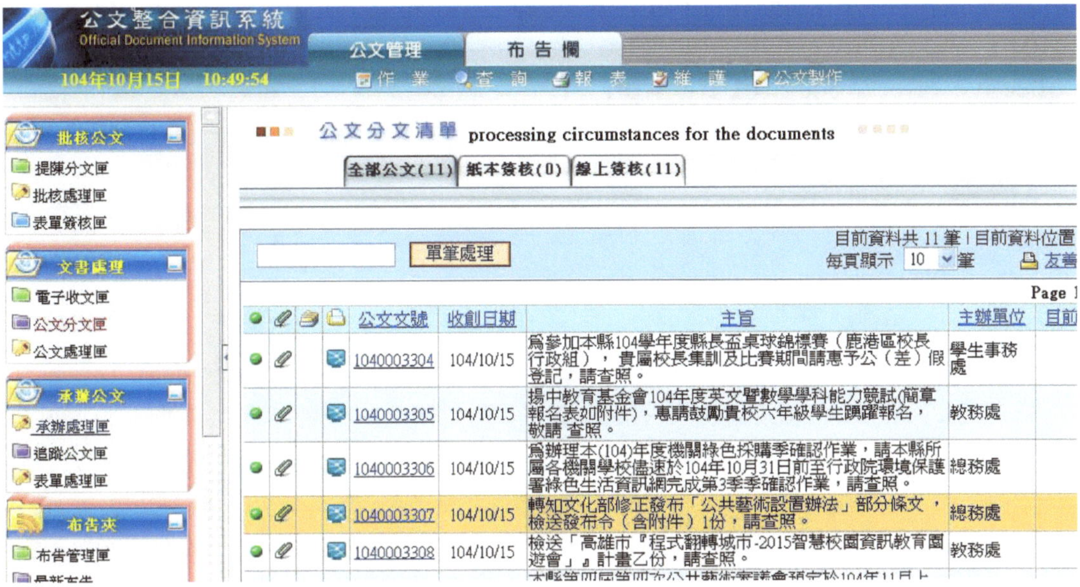

Fig. (2). The supervisor can understand the processing circumstances for the documents in the ODIS.

Second, in addition to the quick function of processing the documents, ODIS also promotes effectively administrative functions in the school. For example, substitute staff must carry out work when the designated staff takes a leave of absence from school. Thus, school tasks must be completed on time by the substitute.

Third, the function for announcements is also applied in this system; it provides opportunities for communication with each other about various thoughts on the documents. This can lead to group meetings to enable staff to discuss the views about the work that need to be communicated (Fig. **3**).

Relationship between ODIS and Distributed Leadership

In distributed leadership, situations are very important factors that interact with leaders and followers (Spillane, 2006). Specifically, from such a perspective,

ODIS as an artifact is categorized to the situation, which is one of three elements in distributed leadership practice. First, leaders and followers interact with ODIS to understand their cognitions and actions agencies as well as further descriptions to truly create the practice of distributed leadership, even quickly and effectively promoting the working function. In other words, it not only affects what the leaders and followers do, but also changes their cognitions and actions and collective effectiveness of the work.

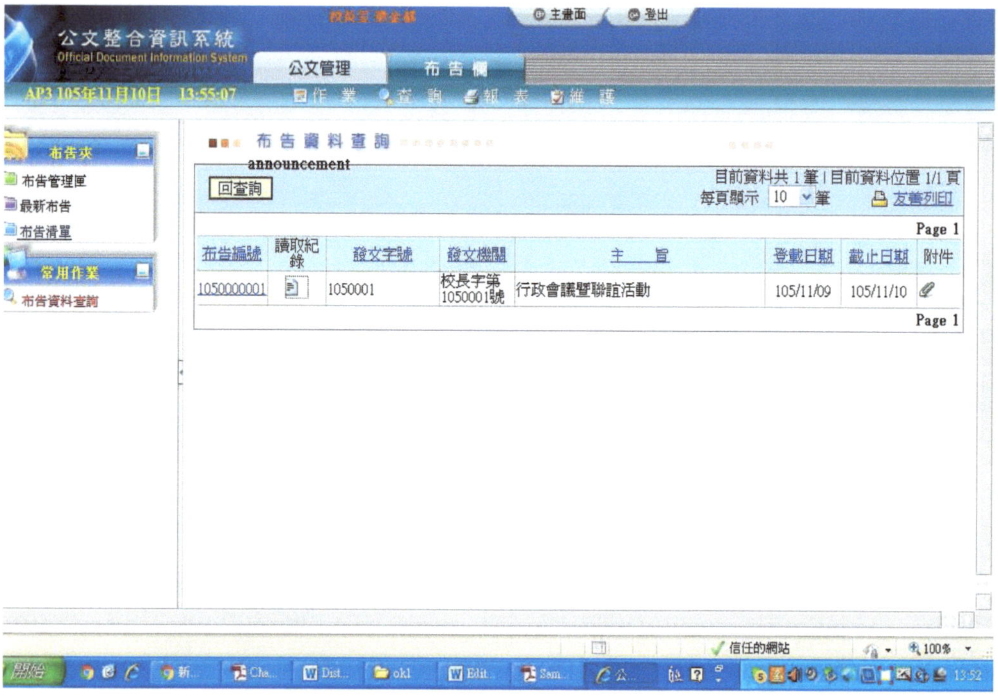

Fig. (3). The function for announcements in ODIS.

In addition, the announcement functions and group meetings *via* ODIS can open new contexts of leadership practice. For example, it may be possible to decide the director of academic affairs as a leader and the principal as a follower in the teaching plan, thereby further constituting a new leadership practice in the school. Thus, ODIS as a situation in distributed leadership in school interacts with leaders and followers, providing a foundation for staff members' professional experiences and further affecting others' cognitions, learning, and development (Lima, 2008) to promote or limit the outcomes of their leading activities in school (Spillane *et al.*, 2004). Based on these views, ODIS as a situation means the staff members' professional abilities can affect others' cognitions and development and collective effectiveness through interactions between leaders and followers.

RESEARCH DESIGN

The author discusses the research design based on the information discussed in the introduction and literature review.

Frame and Methods of the Research

Many studies have suggested the need for case study research that is more comprehensive than many typical studies in educational leadership. As the focus of inquiry expands to include the thinking and actions of multiple actors, and as a context becomes a more important object of study, it becomes necessary to capture a more complex slice of reality (Riehl & Firestone, 2005). In other words, a case study can be used to analyze the availability of the distributed leadership practice in an elementary school because it has the thinking and actions of multiple actors and abundant contexts. Consequently, the author tried to collect data on the interaction of leaders, followers, and situation from August 2013 to August 2015 so as to understand the practice of distributed leadership over time (Spillane, 2006; Spillane *et al.*, 2004). In this study, distributed leadership focuses on the application of ODIS in the Sun Elementary School (Fig. **4**).

Participants and Procedure

For this case study, data were obtained from "principal P" and ten staff members at "Sun Elementary School" (anonymity). There are twenty classes and approximately forty staff members. In the past two years, it started to promote the ODIS to replace the official paper documents, which could change or adjust staff members' working processes and behaviors. Principal P has been a principal for 15 years, possessing good interpersonal relations with staffs, concern for members' needs, and many successful experiences in the school. He has emphasized a positive culture by stressing students' learning, motivating teachers' enthusiasm, promoting teachers' professional abilities, *etc*. The other participants were major administrative staffs who were affiliated with the administrative work for handling the ODIS, such as a director or team leader of academic affairs or counseling students.

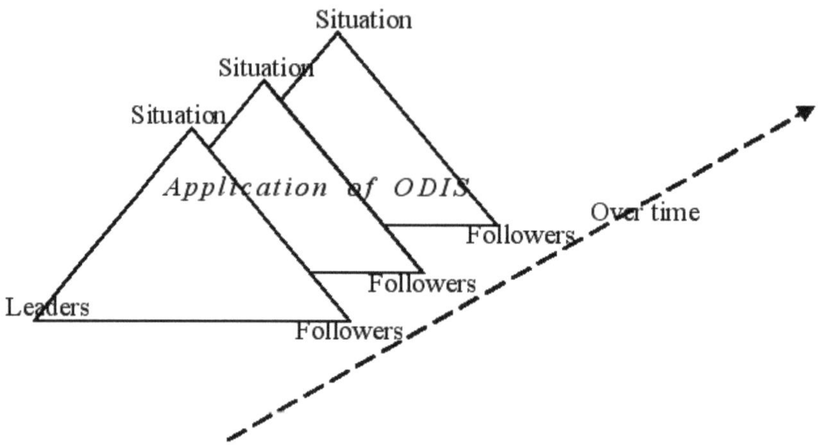

Fig. (4). An analysis the distributed leadership practices in an elementary school through the application of the official document information system.

Note: One triangle represents the single leadership practice, and three triangles stand for three practices. Similarly, the dotted line indicates multiple leadership practices over time, constituting distributed leadership practice.

In addition, participatory observations, interviews, and documents were used in this case study from August 2013 to August 2015 in order to analyze the actions of school members (including the principal and staff members) who interacted with ODIS, thereby understanding the represented meaning (Riehl & Firestone, 2005) as the formed leadership practice. The participatory observations and interviews mainly took place in the office of the principal, the office of school staffs, and the classrooms respectively and focused on the behaviors and actions of principal and staff members. The interviews used an interview outline covering ten topics on ODIS operations; the outline was revised by three scholars and four elementary school principals based on their professional opinions. Furthermore, the contents of documents (*e.g.*, meetings records and official documents) were collected to understand administrative operations in the day-to- day running of the school. Simply speaking, such methods could obtain the correct and proper data by way of a scientific procedure in order to reach the study's purpose.

Concerning these collective methods, the participatory observations were marked "Observation Year-Date" (*e.g.*, Observation 13-1205, Observation 14-0103) and interviews were marked "Interview Year-Date" (*e.g.*, Interview 13-1202, Interview 14-1212). Documents retrieved from the meetings and reflective notes were marked "Document Order" (*e.g.*, Dm 01, Dr 13).

FINDINGS AND DISCUSSIONS

According to the purpose and design of the research conducted in Sun Elementary School from August 2013 to August 2015, the single leadership practice and its evolving multiple practices over time were described through interactions among situations, leaders, and followers in the administrative operations, which took the OIDS as an investigative focal point. They were as follows:

Clearing Up and Down of the Situation←→Responsibility to Oneself and Leaders←→Followers of Assistance, Assignment, and Resistance

Relative to the greater convenience and quality of ODIS, the official paper document processes at Sun Elementary School classically created a "clearing up and down of the situation" (Law *et al.*, 2010), as the official paper document contents were first reviewed by the principal and then assigned to the staff members to take the appropriate actions. In this situation, the principal as the power distributor controls and assigns the work to the staff to practice his own beliefs in school. Such processes might have faced some disagreements and discontent, but they appeared to maintain smooth operations in school affairs over time and completion of the school tasks, as was explained by the director of academic affairs: "Although the principal reviewed and assigned entirely the official documents in school, the responsibilities were often unclear. That is, he assigned them based on his own experiences and opinions… I still did my best to achieve the goal" (Interview 13-0819). The system of leadership practices was based on interactions between staff members and official paper documents and was gradually transformed into school routines (Spillane & Diamond, 2007), which highlighted positions for school members as distinct criteria between leaders and followers in the official distributed pattern of distributed leadership (MacBeath, Oduro, & Waterhouse, 2004).

These situations indicated that staff members' efforts were mostly the layer of directors and team leaders that had been carried out over a period of time. They thoroughly understood the operations for the documents and took their own actions. For example, some could quickly complete the tasks based on their extensive expertise and positive attitude; furthermore, because they did not want to communicate with or feared rejection from others, they solely achieved the goals of the tasks when pertaining to others' affairs with the documents. It also created a bureaucratic style and evasive attitude when some staff only cared about their own needs in taking documents, as the team leader of academic affairs pointed out: "Relative to matching and communicating others' affairs, I thought I favor responding to my own tasks" (Interview 13-1218).

The data further showed that the leadership styles of formal positions usually focused on distributing power (Hulpia *et al.*, 2010), and the subordinates fulfilled the follower roles of assistance and coordination. This interactive style between leaders and followers likely created a mutual struggle and dispute, leading to formalized leadership actions. It could reduce work accountability, as indicated by some teachers: "We only handed over the works according with wishes of the director of students' affairs when she asked for the students' composition on the Plan of Traffic Safety" (Interview 13-1028). In these situations, the followers only helplessly accepted the command and instruction on tasks from leaders as a formal position. It was just a distributive form of person-plus (Perkins, 1993), with few considerations for the interactive relationships among school members, and created counterproductive accountability in the school (Law *et al.*, 2010).

On the whole, in this stage, the leadership style "leaders' individual opinions and abilities" was constituted in the interaction on clearing from up to down the situation, responsibilities of oneself and leaders respectively, and followers of assistance, assignment, and resistance.

Situations of Unstable and Non-Face-to-Face Communicative System←→ Leaders Possessing Expertise and Abilities←→Followers Willing to be Inspected, be Understood, and Interact

In the age of science and technology, school's efforts to reduce wasted resources (*e.g.*, paper and stamps) seem to take place by adapting ODIS rather than executing official paper documents, as requested by the county government (higher authority). However, this unexpected policy brought about some troubles for school staff because they did not possess the expertise for handling the ODIS processes as professional public servants (*e.g.*, director of human resource affairs). In contrast, most staff members were teachers and concurrently directors or team leaders did not handle the ODIS due to their lack of expertise about such processes. Moreover, ODIS appeared the lower accountability in administrative work because of the unstable system and its bad situations. More specially, it emphasized the informative function of the internet, but ignored adoption of the actions in face-to-face communication with each other, as indicated by the director of student affairs:

> The malfunction of the computer system happened frequently one day, so that we opened and closed it over and over. It was very inconvenient…I often talked with the director of academic affairs for one or two minutes and we understood each other's opinions in the past. Nowadays, it was odd. We communicated with each other through the system, which meant few opportunities for face-to-face communication (Interview 14-0303).

For such reasons, the three staff members in charge of this task assigned by the school in such a structure were responsible for receiving, assigning, and issuing the documents as well as checking and maintaining system. Thus, it appeared that one characteristic of distributed leadership was when the added leaders took responsibility for their own work (Lima, 2008). Specially, these staff members as leaders possessed the expertise and abilities related to ODIS, as was described by a reflective note: "The team leader of general affairs was a comprehensive skill for the system. Most members asked her when they had the troubles…she entirely provided assistance and solved their problems again and again" (Dr 05).

In addition, the leaders could deeply understand the followers' roles, what they did, and how they interacted with others (Gunter, 2003). They further considered and analyzed the real circumstances and planned subsequent work schedules and the expected effectiveness in the future. These actors' interactions constituted the cornerstone of organizational leadership (Spillane, 2006). More specially, due to proficiencies with ODIS, the actors started to inform the interrelated departments about their work contents in order to obtain their assistance and achieve the work goals. In the interactive mechanism of the system, followers were willing to be inspected and understood and gradually completed their tasks through their interdependent and collaborative working processes in school (Spillane, 2006). The documented meeting records showed this:

> In ODIS, the principal and team leader of general affairs who were responsible for ODIS, they could control the processes and speed about the documents. Especially, the principal asked for a rigorous deadline on the documents to promote the work effectiveness; for example, a document on affairs of students' scholarship could be completed document as soon as three days (Dm 09).

Specifically, in this stage, the system was unstable, and the leaders possessed the expertise and responsible attitude to encourage the followers to accept the inspection, understanding, and interaction by leaders. It created a leadership practice of "mutual interdependence and interaction". In brief, a productive distributed leadership could integrate the availability of human and material resources from various departments to promote work effectiveness in school (Lai, 2010).

Positive and Responsible Followers←→Situations with Stable System and Additional Multiple-Functions←→Encouraging Mutual Sharing among Leaders

The perspective of distributed cognition stressed that individual cognitions repeatedly changed through the constantly interactive processes between actors and artifacts (Halverson & Clifford, 2006). After operating ODIS proficiently, followers gradually became interested in it and took a step to strengthen their work abilities and adjust their work attitudes. They started to execute this work actively and promote its accountability, as an observation noted: "In this system the school processed one document for three days; it exceeded greatly seven days of the school asked criterion" (Observation 14-1006).

In addition, under the stable and effective system during such a time, they progressively found other additional and positive functions of it, such as the informative internet meetings, unlimited work area system, and substitute position function. These functions not only brought up shares and interactions about opinions on the documents among staff members, but also promoted the entire working effectiveness. With it, the interaction between actors and artifacts incidentally spurred the exchange within actors' views in the school and indirectly elevated individual and collective abilities (Salomon, 1993). As the principal said: "To collect the opinions from the other departments, the informative internet meetings were constantly issued from responsible staffs in school…the substitute function I just set up at the computer… the director of academic affairs very quickly completed the task in ODIS" (Interview 14-1111).

Briefly, the multiple interactions of leaders and followers based on their operation of ODIS seemed to begin across the positional boundaries. Such operations not only encouraged multiple voices from others, but also merged them to refine opinions about the job. Similarly, they created collective cognitions by way of the interaction of internal and external structures on individual staff (Hutchins, 1990) and moved forward a single step to embody the picture of multiple leaders' team (BALOĞLU, 2012; Lima, 2008). That is, the leaders were encouraged by other leaders and followers to engage in more sharing and interactions in this system, as was explained by a team leader of academic affairs: "The internet meetings were adopted many times; especially the Disaster Prevention Plan was taken with staffs from different departments, and second, they were distributed work about it… and it agglomerated the opinions in school and promoted work effectiveness" (Interview 14-1212).

In sum, Spillane and Diamond (2007) pointed out that the expertise of multiple leaders could enhance individual leaders' skills, and the characteristics of collective cognition could also surpass the collections of individuals. It can act as outside agents and inside agents based on the participative discussion process among multiple members (Lima, 2008)—that is, the expertise and experiences of individual and collective members were promoted due to such processes. In short, in the interaction on the positive and responsible followers, situations with a stable system and additional multiple functions as well as encouragement of leaders' mutual sharing constructed the leadership practice of "gradual positive sharing".

Positive Intervention Leaders←→Situation on Research and Development System←→Followers with Expressed Cognitions and Opinions

In addition to promoting the organizational operations, leaders who were in the relevant positions ought to apply their power (Lumby, 2013). Similarly, the Sun Elementary School principal actively pushed internet meetings on ODIS and rewarded staff members for their actions. This not only "formed an atmosphere to support experience sharing and teamwork" (BALOĞLU, 2012), but also upgraded the school's effectiveness, as a director of the counseling room explained:

> About the effectiveness of all dimensions on the system (ODIS) in school, it obtained 100 scores. Its performance was quite perfect. With it, everyone in the participative system got three awards. So the principal took a meal and replaced the new computers to recognize and indicate his gratitude for their performances (Interview 15-0305).

Due to the principal's encouragement and recognition, through constant discussions, two staff members together researched and developed a strong function system to design basic forms as sample documents to share with other institutes, such as schools or the higher authority. In other words, when the responsible staff typed the critical contents into the system, an official document was formed and completed quickly. It could even immediately suggest and revise the vocabulary or writing forms when creating the document.

The supervisor later directly revised the document after reviewing it to promote its qualities. In sum, the available interaction, by way of the ODIS artifact, could advance opportunities for cognition sharing among staff and achieve organizational tasks (Halverson & Clifford, 2006), thereby leading to a step to develop expert leadership (Lima, 2008). As two staff members said separately: "For me, the writing document created bad emotions in the past... They (two staff members) could cooperate to develop such a strong system; it greatly upgraded the effectiveness of these document affairs" (Interview 15-0317). "I could not write the document before these forms. Currently, I just type a few words and it is completed" (Interview 15-0527).

Furthermore, the followers started to express cognitions and opinions on it over time. They not only provided the exchange and brainstormed on working knowledge among staff, but also developed largely to discuss professional teaching knowledge. These development situations gradually confirmed the value of schooling. Namely, followers could become leaders and leaders could become followers, and they can all respond to the common briefs, values, and commitments (Sergiovanni, 2006).

> The director of student affairs explained: "Concerning the teaching ideas, I mostly delivered the views to the system, such as how the remedial teaching system operated on it or how the teaching material was selected...we may exchange opinions with each other, and then put forth creative teaching actions" (Interview 15-0415).

From this, it advanced the professional development of individual and collective teachers when the schools arranged more activities on expertise dialog, which accumulates more social capital for staff through the sharing and discussing of knowledge and experiences (Hands, 2010). Teachers also played the role of various leaders in their professional affairs in order to embody the concept of sharing decisions (Danielson, 2006).

For these reasons, the flat power relationships among staff were derived from the situations of mutual sharing and restructuring in schools (Law *et al.*, 2010) and depicted a picture of participative decision and multiple development (Hulpia *et al.*, 2010). In brief, from the document system functions and the discussion of issues on teaching affairs, it broke the original purpose of ODIS and doubled the effectiveness of the administration and teaching. In this stage, the interaction of positive intervention leaders, situation on research and development system, and followers with expressed cognitions and opinions reflected the leadership practice on "forging research and creativity".

As such, the practice processes on ODIS at Sun Elementary School created the four distributed leadership practice and moved slightly toward productive distributed leadership (Fig. **5**).

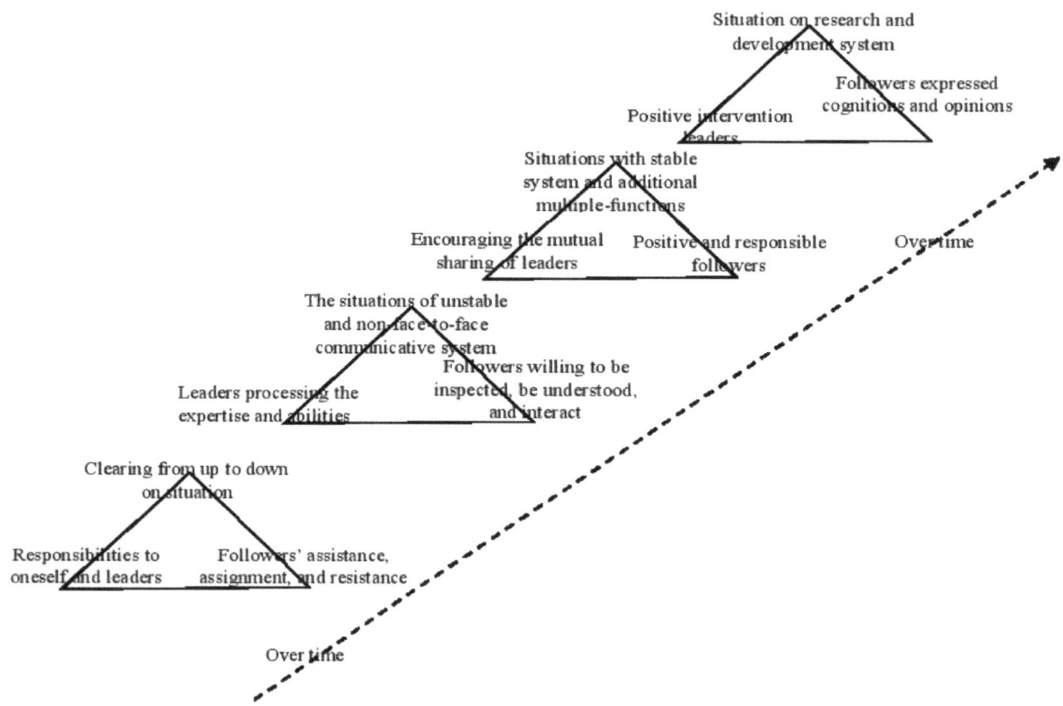

Fig. (5). The processes of distributed leadership practice focused on official document information system.

CONCLUSION

The purpose of this study was to analyze the distributed leadership practices in an elementary school and focus on the application of an ODIS. After analyzing and discussing the data, the conclusion as follows:

Four Leadership Practices: Leaders' Individual Opinions and Abilities, Mutual Interdependence and Interaction, Gradual Positive Sharing, and Forging of Research and Creativity

The findings in this study identified four leadership practices. First, the leadership style on "leaders' individual opinions and abilities" comprises interactions in clearing from up to down on the situation, responsibilities to oneself and leaders, and followers' assistance, assignment, and resistance. Second, in the interaction among the situations of an unstable and non-face-to-face communicative system, leaders possessed the expertise and a responsible attitude, and followers were willing to be inspected by, be understood by, and interact with leaders. This created the leadership practice of "mutual interdependence and interaction". Moreover, the interaction of the situations with the stable system and additional multiple functions, encouraging mutual sharing among leaders and positive and responsible followers, constructed a leadership practice of "gradual positive sharing". Lastly, the interaction of the situation on the research and development system, positive intervention of leaders, and followers with expressed cognitions and opinions it reflected the leadership practice of "forging research and creativity".

Developmental Processes of Distributed Leadership Practice Showed Slight Movement toward a Professional Creative System

The four leadership practices at Sun Elementary School created processes from clearing from up to down on situation, situations of an unstable and non-face-to-face communicative system, and situations with a stable system and additional multiple functions into a situation of a research and development system. The practices transformed the positions and powers, constructing an informational system into a stronger, more researched and developed system. The processes from the center of the person to the integration of the person and science technology moved slightly toward the "situation of a professional creative system".

Distributed Leadership Practice Gradually Created Leaders with Positive Intervention Characteristics

The four leadership practices in this school led to responsibility to oneself and leaders, leaders with expertise and abilities, encouragement of mutual sharing among leaders, and positive intervention leaders. In other words, it transformed the leaders from passive actors facing the affairs, valuing the profession and interactions, into positive interventionists. Thus, the practice with ODIS gradually created leaders with positive intervention characteristics.

Followers Embodied the Driving Force from Passive to Active on the Distributed Leadership

In this research, the distributed leadership in the school created the processes by which followers requiring assistance, assignment, and resistance became willing to accept checks and interactions, be positive and responsible, and expressed their cognitions and opinions. In this sense, the followers gradually transformed the assisted and inspected roles into the expressed cognitions and opinions, thereby embodying the driving force from passive to active in the school.

Entire Distributed Leadership Showed Productive Traits

The school created four leadership practices to guide individuals' opinions and abilities, mutual interdependence and interaction, gradual positive sharing, and forging of research and creativity. The situations shifted from being centered on the person to the integration of the person and science technology, from passively facing affairs to positive intervention leaders, and from passive to active followers. The single trait of moving from being counterproductive to being productive distributed leadership in this school. Thus, the entire distributed leadership in the school showed productive traits.

ODIS was a Key Factor for the Practice of Distributed Leadership Over Time

This study showed that ODIS was constantly interacting with the leaders and followers to create the four leadership practices in the school. In the initial stage, the actors, both leaders and followers, resisted and opposed ODIS, claiming it resulted in inconveniences and environmental changes. Such interactions created new leadership practices as actors had to deal with ODIS in their work environment over and over. Over time, the actors gradually accepted and applied it, promoting their working effectiveness and further liking it when more functions of artifacts were found. In such processes of distributed leadership practices, the key factor of ODIS was evident everywhere in the school.

CONFLICT OF INTEREST

The author confirms that the author has no conflict of interest to declare for this publication.

ACKNOWLEDGEMENTS

Declared none.

REFERENCES

BALOĞLU, N. (2012). Relations between value-based leadership and distributed leadership: A casual research on school principles' behaviors. *Educational Sciences: Theory & Practice, 12*(2), 1375-1378.

Chau, B.I. (2005). File security management for government electrical document. *Government Resource Planning, 29*(1), 46-52.

Danielson, C. (2006). *Teacher leadership that strengthens professional practice.* Alexandria, VA: Association for Supervision and Curriculum Development.

Gronn, P. (2002). Distributed leadership. In: Liethwood, K., Hallinger, P., (Eds.), *Second international handbook of educational leadership and administration.* (pp. 653-696). Dordrecht, Netherlands: Kluwer. [http://dx.doi.org/10.1007/978-94-010-0375-9_23]

Gunter, H. (2003). Introduction: The challenge of distributed leadership. *School Leadership Management, 23*(3), 261-265. [http://dx.doi.org/10.1080/1363243032000112775]

Halverson, R.R., Clifford, M.A. (2006). Evaluation in the wild: A distributed cognition perspective on teacher assessment. *Educational Administration Quarterly, 42*(4), 578-619. [http://dx.doi.org/10.1177/0013161X05285986]

Hands, C.M. (2010). Why collaborate? The differing reasons for secondary school educations' establishment of school-community partnerships. *School Effectiveness and School Improvement, 21*(2), 189-207. [http://dx.doi.org/10.1080/09243450903553993]

Harris, A. (2005). Leading or misleading? Distributed leadership and school improvement. *Journal of*

Curriculum Studies, 37(3), 255-265.
[http://dx.doi.org/10.1080/00220270500038602]

Hulpia, H., Devos, G., Keer, H.V. (2010). The influence of distributed leadership on teachers' organizational commitment: A multilevel approach. *The Journal of Educational Research, 103*, 40-52.
[http://dx.doi.org/10.1080/00220670903231201]

Hulpia, H., Devos, G., Rosseel, Y. (2009). The relationship between the perception of distributed leadership in secondary schools and teachers' and teacher leaders' job satisfaction and organizational commitment. *School Effectiveness and School Improvement, 20*(3), 291-317.
[http://dx.doi.org/10.1080/09243450902909840]

Hutchins, E. (1990). The social organization of distributed cognition. In: Resnick, L.B., Levine, J.M., Teasey, S.D., (Eds.), *Perspectives on Socially Shared Cognition* (pp. 283-307). Washington, DC: American Psychological Association.

Hutchins, E. (1995). How a cockpit remembers its speed. *Cognitive Science, 19*, 265-288.
[http://dx.doi.org/10.1207/s15516709cog1903_1]

Lai, C.F. (2010). *Current issues in school leadership*. Taipei, TW.

Law, E., Galton, M., Wan, S. (2010). Distributed curriculum leadership in action: A Hong Kong case study. *Educational Management Administration and Leadership, 38*(3), 286-303.
[http://dx.doi.org/10.1177/1741143209359714]

Leithwood, K., Duke, D.L. (1999). A century's quest to understand school leadership. In: Murphy, J., Louis, K.S., (Eds.), *Handbook of Research on Educational Administration* (pp. 45-72). San Francisco, CA: Jossey-Bass.

Lima, J.A. (2008). Department networks and distributed leadership in schools. *School of Leadership Management, 28*(2), 159-187.
[http://dx.doi.org/10.1080/13632430801969864]

Lloyd, S. (2005). *Distributed leadership: The rhetoric and the reality.* London, UK: Forest Gate Community School.

Lumby, J. (2013). Distributed leadership: The uses and abuses of power. *Educational Management Administration and Leadership, 41*(5), 581-597.
[http://dx.doi.org/10.1177/1741143213489288]

MacBeath, J., Oduro, G., Waterhouse, J. (2004). *Distributed leadership in action: Full report.* Nottingham, UK: National College for School Leadership.

Perkins, D.N. (1993). Person-plus: A distributed view of thinking and learning. In: Salomon, G., (Ed.), *Distributed cognition: Psychological and Educational Considerations* (pp. 88-110). New York, NY: Cambridge University.

Riehl, C., Firestone, W.A. (2005). What research methods should be used to study educational leadership? In:

Firestone, W.A., Riehl, C., (Eds.), *A new agenda for research in educational leadership.* (pp. 156-170). New York, NY: Teachers College.

Salomon, G. (1993). No distribution without individuals' cognition. In: Salomon, G., (Ed.), *Distributed cognition: Psychological and Educational Considerations* (pp. 111-138). New York, NY: Cambridge University.

Sergiovanni, T.J. (2006). *The principalship: A reflective practice perspective.* (5th ed.). Boston, MA: Allyn & Bacon.

Spillane, J.P. (2005). Distributed leadership. *The Educational Forum, 69*(2), 143-150. [http://dx.doi.org/10.1080/00131720508984678]

Spillane, J.P. (2006). *Distributed leadership.* San Francisco, CA: Jossey-Bass.

Spillane, J.P., Diamond, J.B. (2007). Taking a distributed perspective. In: Spillane, J.P., Diamond, J.B., (Eds.), *Distributed leadership in practice.* (pp. 1-15). New York, NY: Teachers College.

Spillane, J.P., Halverson, R., Diamond, J.B. (2004). Towards a theory of leadership practice: A distributed perspective. *Journal of Curriculum Studies, 30*(1), 3-34. [http://dx.doi.org/10.1080/0022027032000106726]

Timperley, H.S. (2005). Distributed leadership: Developing from practice. *Journal of Curriculum Studies, 37*(4), 395-420. [http://dx.doi.org/10.1080/00220270500038545]

Tsai, S.T. (2010). Current situation and prospects on electrical document for carbon reduction. *Government Resource Planning, 34*(4), 73-78.

Yekovich, F.R. (1993). A theoretical view of the development of expertise in credit administration. In: Hallinger, P., Leithwood, K., Murphy, J., (Eds.), *Cognitive perspectives on educational leadership* (pp. 146-166). New York, NY: Teachers College.

CHAPTER 8

E-Knowledge Consortium Shikoku: E-Learning Activities on Local University Alliance

Toshihiro Hayashi*

Information Technology Center, Kagawa University, Takamatsu, Kagawa, Japan

Abstract: *Shikoku* is an island of Japan. To create cooperative region where the people can share the awareness of "*Shikoku* is one community", eight universities in *Shikoku* established e-Knowledge Consortium Shikoku (eK4) for realizing educational collaboration by e-Learning for the purpose of human resource development for *Shikoku* region. The consortium organizes e-Learning contents of the characteristic lectures of the eight universities in order to provide e-Learning courses among them. The e-Learning infrastructure of eK4 has typical features of distributed LMS and Shibboleth identification. eK4 has provided serval e-Learning courses based on credit transfer system since 2010. This chapter describes eK4 as a local university alliance and the educational activities in the viewpoint of providing e-Learning courses.

Keywords: Chipla-e project, Distributed LMS, e-Knowledge Consortium Shikoku, e-Learning, eK4, Human resource development, Local university alliance, Shibboleth, *Shikoku*, Study of *Shikoku*.

INTRODUCTION

Shikoku --- An Island of Japan

Japan consists of four main islands (*Honshu*, *Kyushu*, *Shikoku* and *Hokkaido*) and a lot of small islands. *Shikoku* is the smallest in the main islands. Fig. (**1**) shows the main islands of Japan and magnified *Shikoku* island map. *Shikoku* is located in

* **Corresponding author Toshihiro Hayashi:** Information Technology Center, Kagawa University, Takamatsu, Kagawa, Japan; Tel/Fax: +81-87-864-2227/2262; E-mail: hayashi@eng.kagawa-u.ac.jp

Kuo Hung Huang (Ed.)

the south-west side of Japan so that the climate of *Shikoku* is generally warm. It can be said *Shikoku* is easy for people to live.

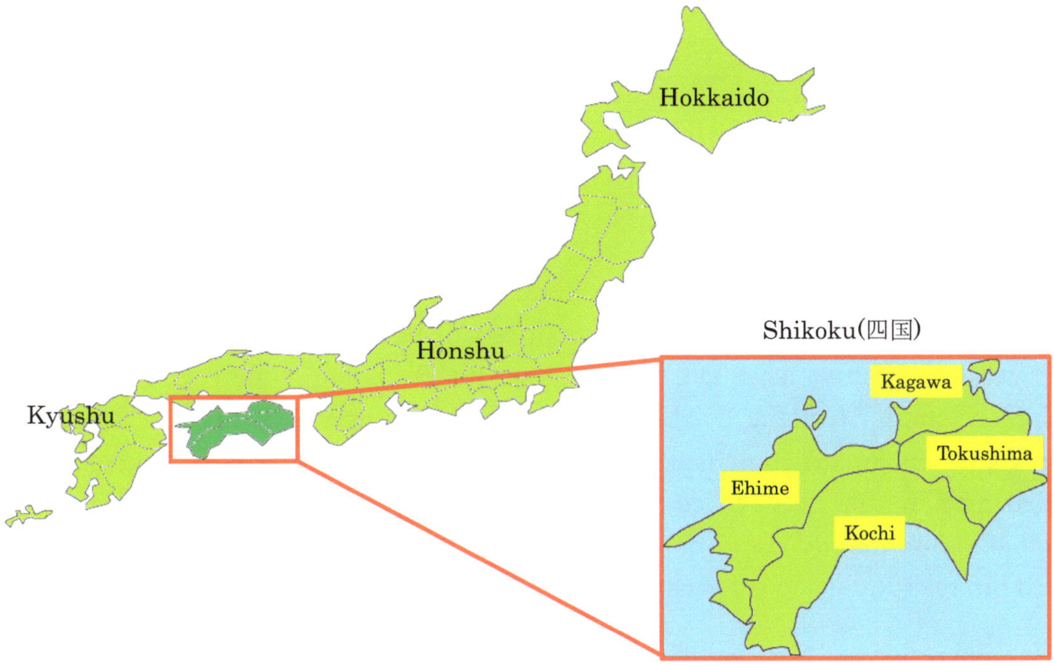

Fig. (1). *Shikoku* in Japan and the four prefectures.

Shikoku is consists of four prefectures: *Kagawa*, *Tokushima*, *Ehime* and *Kochi*. This administrative structure is derived from four old provinces in *Shikoku* respectively *Sanuki*, *Awa*, *Iyo* and *Tosa*. The name of *Shikoku* means four provinces. Four provinces originally had different history, geography, products, culture and tradition. These features have been inherited to current four prefectures in *Shikoku*.

As the total performance of four prefectures, current economical scale of *Shikoku* is small compared with other areas in Japan. Young people tend to leave from *Shikoku* for getting their jobs and for having convenient life style at urban areas. Therefore, *Shikoku* is now facing serious issues such as low birthrate and longevity as a worse result of the population flow. In order to solve these serious issues, talented persons who have high specialty and think of *Shikoku* region are needed. This is a substantial background of e-Learning activities on local university alliance and collaboration in *Shikoku*.

University Collaborations in Japan

Recently, there have been increased new styles of educational practices based on collaboration by plural higher educational organizations including universities. Japanese government has also enhanced such university collaborations by special budgets for their activities and so on. The followings are concrete examples of university collaborations.

University Consortium Saga (University Consortium Saga, n.d.) realizes cooperative education by six universities located in Saga prefecture (Saga University, Nishikyushu University, Kyushu Ryukoku Junior College, Saga Women's Junior College, Nishikyushu University Junior College and The Open University of Japan). Saga is a prefecture in Kyusyu island of Japan and is not so urban area. In this cooperative education, Saga University as a national university in this region mainly coordinates various kind of educational activities including credit transfer by using e-Learning.

The Consortium of National Universities in Hokkaido for Liberal Arts Education (The Consortium of National Universities in Hokkaido for Liberal Arts Education, n.d.) consists of seven national universities located in Hokkaido (Hokkaido University, Hokkaido University of Education, Muroran Institute of Technology, Otaru University of Commerce, Obihiro University of Agriculture and Veterinary Medicine, Asahikawa Medical University and Kitami Institute of Technology). This consortium realizes liberal arts education by credit transfer system using bi-directional remote lecture systems.

In addition, Kyoto Institute of Technology, Kyoto Prefectural University and Kyoto Prefectural University of Medicine collaboratively provides same liberal arts education. This liberal arts education is managed by Institute of Liberal Arts and Sciences (Institute of Liberal Arts and Sciences, n.d.). The typical feature is that getting credit is not realized by credit transfer system but also students can take other universities' classes as those of own universities.

Overview of eK4

In 2008, e-Knowledge Consortium Shikoku (called "eK4" hearafter) (e-Knowledge Consortium Shikoku, n.d.) was established for training talented persons who have high specialty and think of *Shikoku* region. eK4 is currently consists of eight universities in *Shikoku* (Tokushima University, Naruto University of Education, Kagawa University, Ehime University, Kochi University, Shikoku University, Tokushima Bunri University and Kochi University of Technology). This consortium treats with educational collaboration by e-Learning based on credit transfer system.

Seven of these universities proposed an educational project named "Human Resource Development for Promoting *Shikoku* Region based on Integration of Knowledge of *Shikoku*" to Support Project for Strategic University Collaborations by MEXT (Mistry of Education, Culture, Sports, Science and Technology) of Japan (Mistry of Education, Culture, Sports, Science and Technology of Japan, n.d.). The proposed project was accepted and then eK4 activities started since 2008. The typical features of e-Learning infrastructure of eK4 are distributed LMSs (Learning Management Systems) and Shibboleth authentication. The eight universities of eK4 have started e-Learning based on credit transfer system since 2010. As the results of academic year of 2014, fourteen e-Learning courses as liberal arts education were provided, and 771 students registered the courses in eK4 universities.

E-KNOWLEDGE CONSORTIUM SHIKOKU

This section describe e-Knowledge Consortium Shikoku from three viewpoints: (1) the purpose of eK4, (2) features of e-Learning course provided by eK4 universities, and (3) how to take e-Learning courses on credit transfer system. The system infrastructure of e-Learning is described in the next chapter.

Purpose of eK4

eK4 is an educational consortium which consists of eight universities in *Shikoku*: five national universities (Tokushima University, Naruto University of Education, Kagawa University, Ehime University and Kochi University), two private universities (Shikoku University and Tokushima Bunri University) and one public universities (Kochi University of Technology). The consortium was established as the main organization to promote the educational project of "Human Resource Development for Promoting *Shikoku* Region based on Integration of Knowledge of *Shikoku*".

The project of eK4 was one of university support projects by MEXT of Japan. Their purpose was to generate "The Base of Knowledge" in respective regions by active cooperation among public and private universities and by effectively utilizing education and research resources in each university. The Base of Knowledge is a key concept of the support projects and eK4 project also has a similar concept called "Knowledge of *Shikoku*". eK4 shares Knowledge of *Shikoku* as the education and research resources of eight universities by using ICT. Fig. (**2**) shows serveral examples of the education and research resources with the characteristics of universities in *Shikoku*.

Fig. (2). Knowledge of *Shikoku*.

The purpose of eK4 is to train talented persons who have high specialty and think of *Shikoku* region. They are required to commit to develop *Shikoku* cooperatively in order to enhance self-supported regional development. Fig. (**3**) shows the concept of human resource development for *Shikoku* region coordinated by eK4. The project tries to develop two types of talented persons: persons who activate *Shikoku* and persons who lead primary industries.

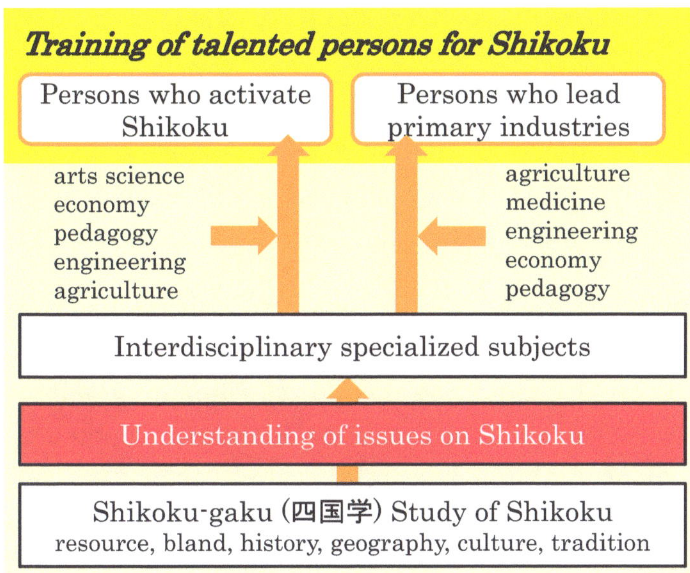

Fig. (3). Training of talented persons who have high specialty and think of *Shikoku* region.

It is very important for such kind of talented persons to understand wide range issues of *Shikoku* region. To enhance understandings of the issues of *Shikoku*, we believe we should provide them two types of subjects as university education: liberal art subjects which treat with *Shikoku* called *Shikoku-gaku* in Japanese and interdisciplinary specialized subjects. The latter subjects include both liberal art subjects and specialized subjects.

Unfortunately, it is hard to provide the above-mentioned subjects to such target students in universities of eK4 all at once by geographical distance. In order to solve this problem, eK4 has adopted e-Learning as delivery way of such subjects to the eight universities. The final goal of eK4 is not to provide e-Learning subjects but also to develop human resource for *Shikoku* region. However, this chapter put focus on mainly the part of educational infrastructure for it.

As another results of eK4 project, we have also expected the following three advantages for *Shikoku* region:

1. The consciousness of "A member of *Shikoku*" arises,
2. "Intellect of *Shikoku*" improves, and
3. "Attractiveness of *Shikoku*" is delivered to the whole country.

Features of eK4 e-Learning Course

In order to train persons who are important for *Shikoku* region, we provide *Shikoku-gaku* (Study of *Shikoku*) subjects which contain educational topics about resource, bland, history, geography, culture, tradition and so on of *Shikoku* region.

We generally bring resources of *Shikoku-gaku* from the results of representative education and researches of eight universities called Knowledge of *Shikoku*. For example, ocean researches about *Seto* inland sea (including *Seto* inland shallow sea) and the *Kuroshio* Current (Japan Current), environmental researches about *Satoyama* and *Satoumi*, cultural researches about history of *Shikoku* pilgrims called *Ohenro* or *Oshikoku-san*, and disaster prevention researches to the giant earthquake called *Nankai-Jishin* (*Nankai* earthquake) are included in Knowledge of *Shikoku*. Because the concept of Knowledge of *Shikoku* actually covers wide arrange topics, we select *Shikoku-gaku* subjects from four categories: history, literary arts, society and nature.

On the other hand, interdisciplinary specialized subjects are defined as higher level subjects for students who have already taken *Sikoku-gaku* subjects. Therefore, the interdisciplinary specialized subjects cover both undergraduate and postgraduate courses (including both liberal art subjects and specialized subjects).

Shikoku-gaku subjects are classified into two styles: omnibus style and regular style. The former style of subjects are constructed from same theme (history, literal arts, society, and nature) e-Learning contents provided by plural universities of eK4. We have three omnibus style e-Learning subjects: "History and Culture of *Shikoku* Area" as a joint subject of history and literal arts, "Local Development of *Shikoku* Area" as a subject of society, and "Natural Environment and Disaster Measures of *Shikoku* Area" as a subject of nature. Fig. (**4**) is a snapshot of e-Learning contents of "History and Culture of *Shikoku* Area". In addition, Table **1** shows representative topics of three *Shikoku-gaku* subjects with omnibus style.

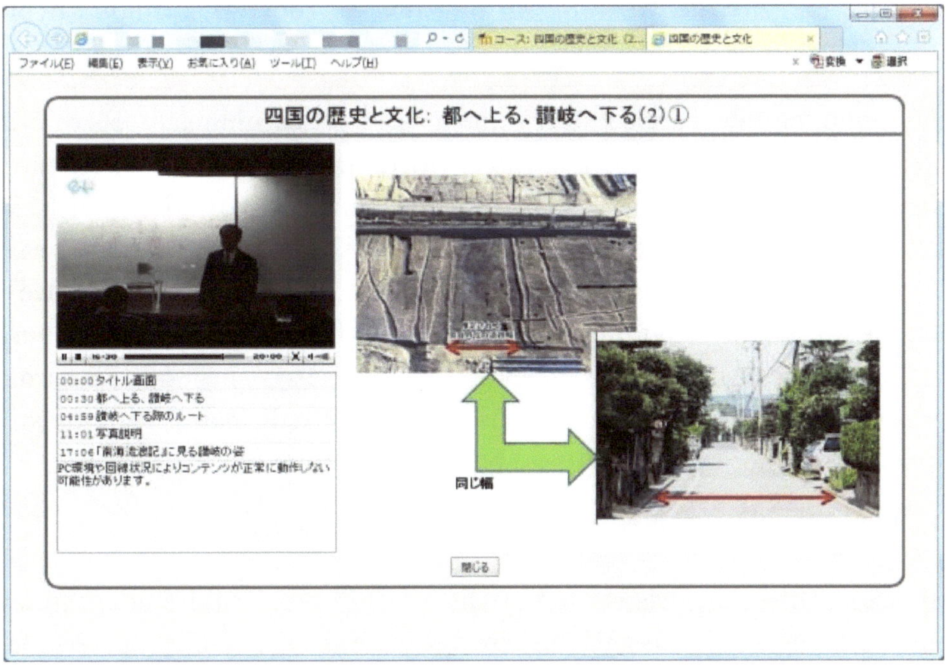

Fig. (4). A snapshot of e-Learning contents.

Table 1. Representative topics of *Shikoku-gaku* subjects.

Subject	**Representative Topics**
History and Culture of Shikoku Area	Powerful thanes in *Sanuki* Interchange among *Awa*, *Edo*, and *Osaka*
Local Development of Shikoku Area	Fish farming in *Kagawa* Cooperation between local industry and university Forestry in *Koch*
Natural Environment and Disaster Measures of Shikoku Area	History of *Tameike* ponds Climate and water resource of *Shikoku* Islands of *Seto* Inland Sea Environmental study of *Kuroshio* current and sea of *Tosa* Making of hazard map for disaster prevention

On the other hand, the regular style of *Shikoku-gaku* is provided as a subject with original theme by one university of K4. For example, Naruto University of Education provides a subject titled "*Awa-gaku*" (Study of *Tokushima*, the sub-title is Local Culture Research) and Tokushima Bunri University provides a subject titled "Local Government Theory". Especially, *Awa-gaku* consists of e-Learning and exercise of *Aruki-henro* (*Shikoku* pilgrims on foot) as a unique hybrid e-Learning subject of eK4.

Currently, both omnibus style and regular style of e-Learning subjects of eK4 are all provided as asynchronous e-Learning (video on-demand style of e-Learning). eK4 tried synchronous e-Learning (live broadcasting style of e-Learning) subject with use of video conference systems in 2010. Unfortunately, it was not easy for instructors and staffs to handle with synchronous e-Learning so that we have completely shifted to providing asynchronous e-Learning subjects in our eK4 projects. However, this does not mean that we refuse synchronous e-Learning in eK4 project. We also think synchronous e-Learning will match with the style of intensive courses (not-regular courses) better than synchronous e-Learning.

How to Take e-Learning Courses and to Get the Credit

In 2010, our consortium has started credit transfer system for student to take e-Learning subjects and to get the credits among the eight universities by exchanging copies of two agreements about credit transfer in *Shikoku* cooperative universities of eK4. The credit transfer system of eK4 is constructed on the model of that for face-to-face subjects. Table **2** shows a typical flow of credit transfer system that a student of *B* University gets the credit of an e-Learning subject of *A* University.

Table 2. Typical flow of credit transfer system in eK4.

1.	**Announcement for registration of credit transfer subjects (at *A* University)**
	1.1 Selection of credit transfer subjects
	1.2 Making of a guidebook for registration of credit transfer subjects of *A* University
	1.3 Sending of the guidebook to cooperative universities including *B* University
2.	**Selection of credit transfer subjects (at *B* University)**
	2.1 Selection of subjects available for students of *B* University to get
	2.2 Announcement of the list of subjects available to get as credit transfer subjects

(Table 2) contd.....

3.	**Taking of credit transfer subjects (at *B* University)**
	3-1 Sending of applications for subject registration to *A* Universities
	3-2 Taking of credit transfer subjects by e-Learning
4.	**Judgement of giving credits of registered subjects (at *A* University)**
	4.1 Examination of registered subjects at *B* University
	4.2 Sending of examination results to *A* University
5.	**Exchange of subjects between *A* University and *B* University (at *B* University)**
	5-1 Acceptance of passed subjects of *A* University as those of *B* University
	5.2 Getting of the credits

Advantages and Disadvantages of Credit Transfer System of eK4

We found both advantages and disadvantages of e-Learning by credit transfer system through educational practices of eK4 in several years. For examples, in order to take e-Learning subjects provided by Kagawa University, all registered students except Kagawa University must become special registration students of Kagawa University. This means all registered students belong to Kagawa University. From the viewpoint of responsible instructors for the e-Learning subjects, they do not take care of difference of each student's affiliation. Most of their educational tasks are similar to as those of regular face-to-face subjects. This is one of the advantages which make easy for instructors to start credit transfer subjects, although e-Learning requires additional tasks such as contents making, mentoring and so on. However, each university of eK4 assists them with such additional tasks by locating support staffs or establishing support organizations.

On the other hand, in the same case of Kagawa University, registration guidance and final examination same as those of face-to-face subjects were required, because the university affairs committee of Kagawa University required the guarantee of educational equality between face-to-face education and e-Learning. Therefore, several devices were needed to realize face-to-face registration guidance and final examination in e-Learning. As for the registration guidance, synchronous remote guidance with video conferencing systems have been given to all students in remote classes. As for final examination, synchronous remote examination have been held at each universities at same time like the university entrance examination given by the National Center of Japan. These operations are troublesome to both students and staffs in the eight universities of eK4. In addition, it was revealed that students were required more procedures to take credit transfer subjects and needed more time to get the credits compared with regular face-to-face subjects. These points become serious barrier for students to take e-Learning subjects.

INFRASTRACTURE OF E-LEARNING

Our consortium has constructed e-Learning infrastructure since 2008, in order that plural universities provide asynchronous and synchronous e-Learning cooperatively. Fig. (**5**) shows the conceptual system infrastructure for e-Learning activities of eK4. The key elements of eL4 infrastructure are distributed LMSs, Shibboleth authentication (Shibboleth, n.d.), and multipoint TV conference systems.

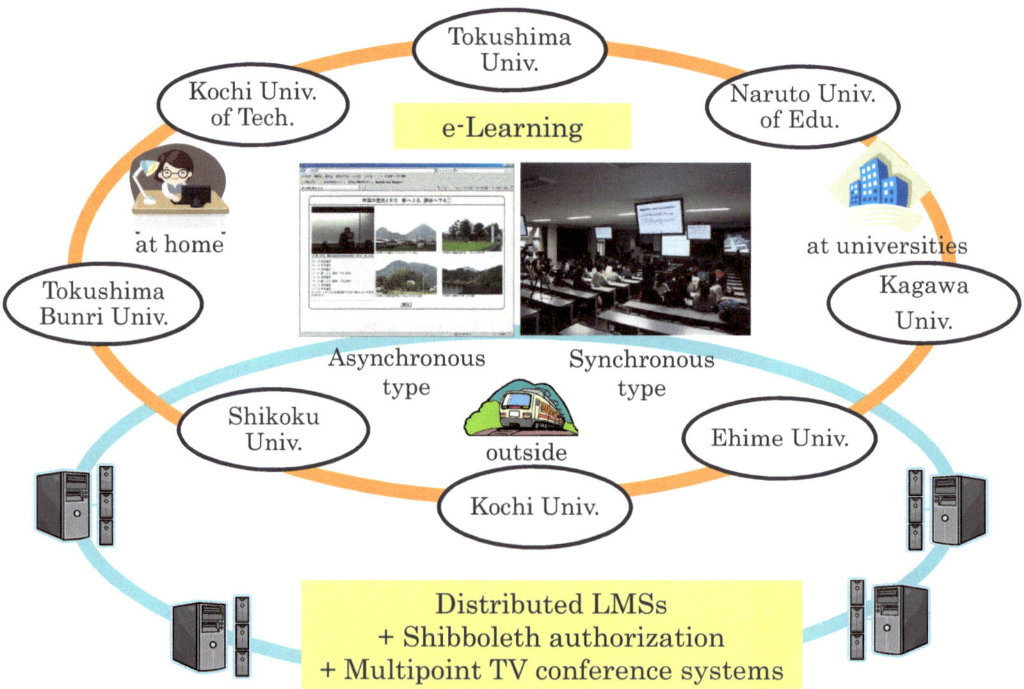

Fig. (5). Conceptual system infrastructure of eK4.

System Infrastructure for Synchronous e-Learning

Our consortium introduced two types of remote conference systems: TV conference systems and web-based conference systems, for realizing synchronous e-Learning. Each university of eK4 introduced both systems in own university. Polycom is representative vender of TV conference systems introduced in eK4 universities, but other conference systems produced by TANDBERG (currently, Cisco Systems) and SONY have been also introduced. We have complicated system infrastructure as multi-vender conference systems for asynchronous e-Learning. In addition, several systems have MCU (Multipoint Commination Unit) functions to realize multipoint connection among eK4 universities. On the other hand, all universities of eK4 introduced Microsoft LiveMeeting as a unified web-based conference systems. LiveMeeting was an online meeting software based on server/client system for realizing real time cooperative works. It provides multipoint remote conference using USB camera, microphone and speakers. LiveMeeting service has finished and changed to Microsoft Lync service. Every university also introduced special communication hardware device called "Microsoft RoundTable" for enhancing LiveMeeting functions. RoundTable has a 360 degree panorama camera constructed from multi-angle cameras. It can close-up a participant who is speaking by using detection function of voice direction and multi-angle cameras, in order to realize online meeting with the sense of reality.

As for a concrete example of conference systems, Kagawa University introduced four TV conference systems produced by Polycom. The one system is HDX 8004 XLP which can handle with high definition images and has MCU function for connecting four points. Other three systems are VSX 6000 which only handles with standard images and has no MCU function. Kagawa University has four campus. HDX 8004 XLP has been located in the main campus and VSX 6000s have been located in other three campuses respectively. On the other hand, Kagawa University introduced five sets of web-based conference system with LiveMeeting and RoundTable. The two sets have been located in the main campus and the three sets have been located in other three campuses. Kagawa University can operate both TV conference and web-based conference among four campuses. Each eK4 university has also introduced remote conference systems for synchronous e-Learning.

Our consortium adopted two ways to realize synchronous e-Learning connected with the eight universities by TV conference systems. One is the use of TV conference systems which have MCU for eight point connection. The other is cascade connection of TV conference systems which have MCU for four point connection. As for the former way, Tokushima University has introduced a TV conference systems which have MCU for connecting eight points. We could realize eight point connection by both ways but we also found some problems to operate it. As for the former way, the base conference system sometimes become unstable in long time operation. As for the latter way, it was hard task to establish eight point connection, because the connection becomes unstable by cascade patterns and other factors such as network status and so on. The conference operator must find optimum (at least "better" even in worst case) cascade pattern. Compared with multi point connection by TV conference systems, it is easy to operate same scale conference by LiveMeeting. However, we-based conference system such as LiveMeeting is not suitable for transmitting high-quality images.

Kagawa University provided a synchronous e-Learning subject about *Shikoku-gaku* in history category in 2010. We found other factors except arrangement of hardware and software become walls to operate synchronous e-Learning subjects. For example, gaps of start and end times prevented from realizing synchronous e-Learning subject on each university's timetable. It was hard to adjust the timetables. In addition, one or more staffs must support synchronous e-Learning subject at every university. We gave up to provide synchronous e-Learning subjects sustainably. However, synchronous e-Learning systems have been used for short-span cooperative e-Learning activities, remote seminars and meetings related to eK4, and so on. "Global Challenge Seminar" shown in Fig. (**6**) is a representative activity which uses synchronous e-Learning systems. This seminar is one-day event brought by Shikoku Economic Federation and eK4, wherein a couple of business persons as instructor give lectures to students of eK4 universities by using the system infrastructure of synchronous e-Learning. In addition, remote seminars and meetings related to eK4 have often held on the synchronous e-Learning environment. Although eK4 universities have shifted their e-Learning to asynchronous style, we do not refuse use of synchronous e-Learning.

Fig. (6). Global challenge seminar.

System Infrastructure for Asynchronous e-Learning

In order to realize asynchronous e-Learning in the eight universities, eK4 constructed a unique e-Learning infrastructure with distributed LMSs (Learning Management Systems) and Shibboleth authentication. Fig. (7) shows the illustrated asynchronous e-Learning infrastructure of eK4 and how to access distributed LMSs. Fig. (7) shows the structure of two universities but eK4 has same system structure in the eight universities. In other words, all universities of eK4 have unified system structure for providing asynchronous e-Learning subjects.

As for management of LMS, we can generally select a centralized LMS or plural distributed LMSs. In the case of eK4, distributed LMSs was adopted for realizing asynchronous e-Learning environment among the eight universities. In other words, the eight universities of eK4 have own LMS respectively. Distributed LMSs have some advantages compared with centralized LMS. For example, instructors of universities that provide e-Learning subjects can store their e-Learning contents on the LMS of own university and can deliver them from the LMSs. Registered students are only own university students and special registration students (they are also own university students). The relation between each LMS and the users is closed in each universities. This means that other university users never use other university's LMS. Therefore, they do not take care of information security policy ruled by each university. In addition, the intellectual property procedure for use of e-Learning contents does not become serious problems, because of using own university's LMS.

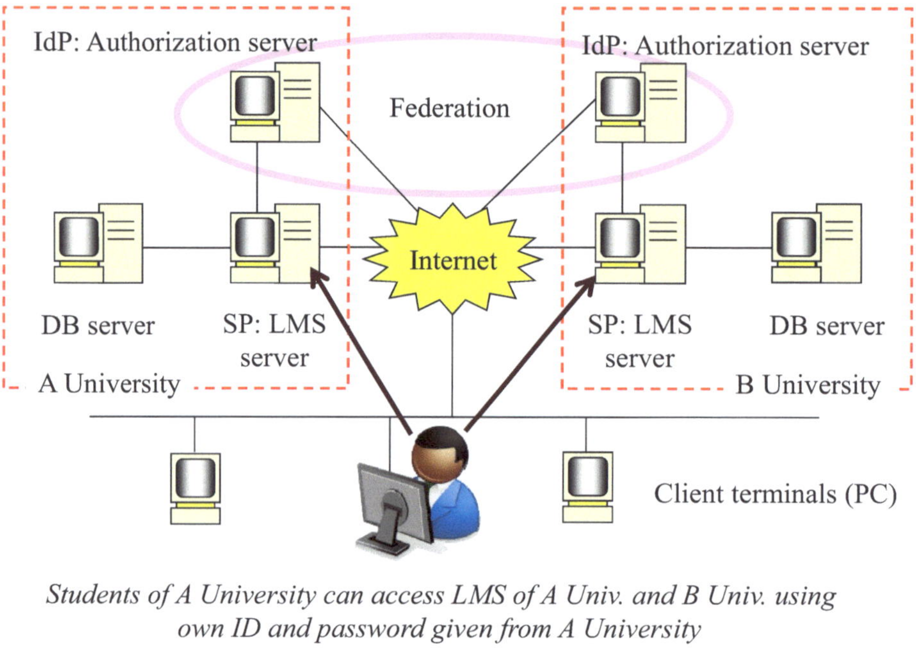

Fig. (7). Distributed LMSs and Shibboleth authentication.

On the other hand, special registration students must access LMSs of other universities. Therefore, it becomes a serious problem how to handle with login ID and the password for access such LMSs. Although the LMS administrator can set local accounts (ID and the password) for them to access the LMS, the administrator or e-Learning support staffs must send the account information quickly and safely to special registration students. This process increases their task loads. In addition, information security policy ruled by each eK4 university requires the staffs to do this process with no mistakes in according to the strict security rules. In order to reduce this task, our consortium has adopted Shibboleth authentication based on eK4 universities federation. To realize Shibboleth authentication environment, each eK4 university prepared an IdP (identity provider) server and Kagawa University, in addition, prepared a DS (Discovery

Service) server. IdP works as user authentication for each university users to access other universities' LMSs. Therefore, each university does not need to handle with account information for other universities students, In addition, such students can access other universities' LMS with their ID and password provided by their own university. However, the students must remember the URL information of their IdP. DS server stores the list of eK4 universities' Idp information in order for the students to reduce such memorization. LMSs work as SP (service provider) get account information on Shibboleth authentication. The Shibboleth authentication infrastructure for eK4 consists of eight sets of IdP and SP, and one DS. This means each eK4 university has own IdP and SP. In addition, all eK4 universities have adopted Moodle as their LMSs (SP).

Fig. (**8**) shows how students can access other eK4 university's LMS. First, they access the target LMS (*step1*). Then, they select shibboleth authentication on the LMS. The web page goes to the DS server shown in Fig. (**9**) (*step2*). By selecting own university from the university list on the page, the web page is moved to the IdP of own university (*step3*). After the user account information is accepted, the web page goes to the target LMS again (*step4*). Finally, the students can get access right of the LMS. Although this process is more complicated than general Moodle login process, students can access any eK4 LMSs by using own ID and password.

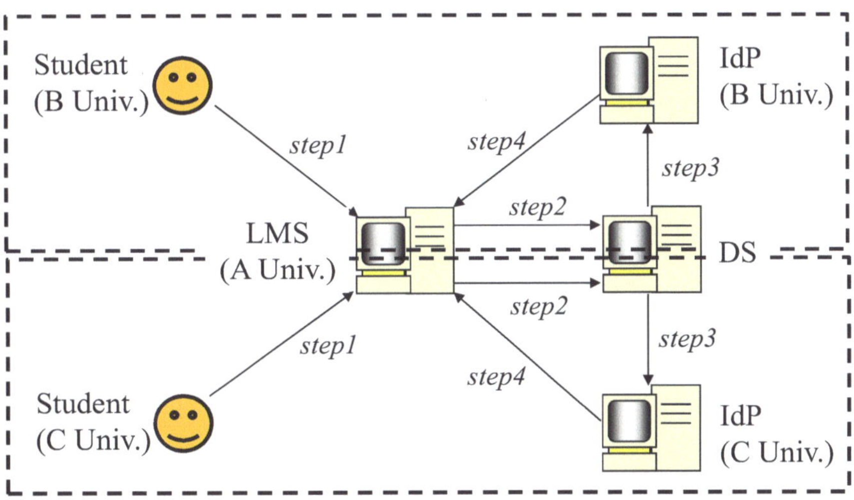

Fig. (8). How to access to other eK4 university's LMS.

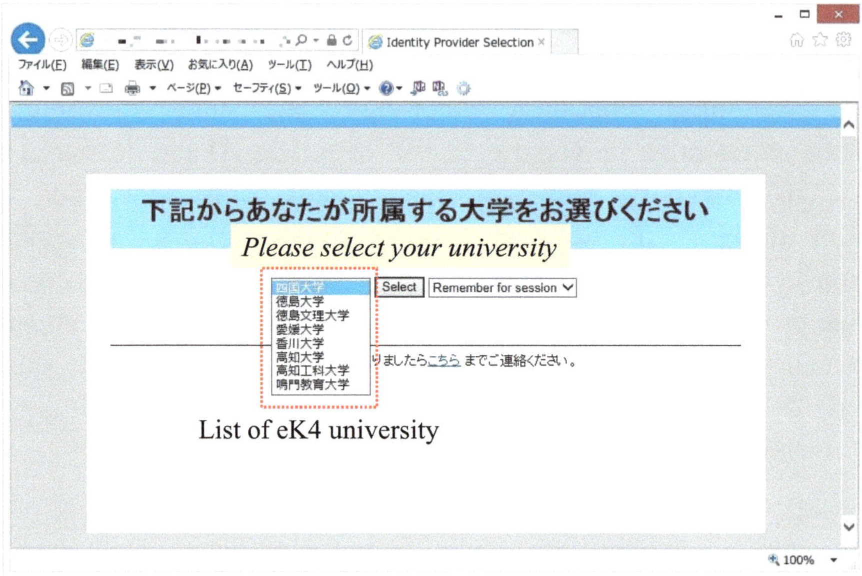

Fig. (9). Interface of DS for eK4.

EK4 ACHIVEMENT

As mentioned above, our consortium has been running the credit transfer system with e-Learning subjects among eight eK4 universities since 2010. This section explains about e-Learning subjects provided in 2013. Then, the results of e-Learning activities in seven years (2008 - 2014) are also described here from the three viewpoints: (1) the number of e-Learning contents, (2) the number of e-Learning courses, and (3) the number of total e-Learning subject registration. In addition, this section explains the issues of eK4 credit transfer system.

eK4 Subjects Provided in 2013

eK4 subjects are classified into two styles: omnibus style and regular style. Table **3** shows three omnibus style subjects provides as eK4 subject provided in 2013. These subjects are all classified into *Shikoku-gaku*. “Subject provider” means a university which provides the subject and “contents provider” means a university which provides one or more e-Learning contents for the subject. In the case of “Local Development of Shikoku Area”, Kagawa University provides nine contents, Kochi University provides two contents, and Tokushima Bunri University provides three contents respectively. However, Kagawa University only provides the subjects. In the viewpoint of students, it is same as other subjects for students of Kagawa University but it is a credit transfer subject for students of other universities. Kagawa University as the responsible organization for eK4 project has provided these three *Shikoku-gaku* subjects since 2010, because they are core subjects of eK4 projects. The numbers in parentheses of "Total" column are the total numbers of each subject registration by students of universities which do not provide the subject. For example, "History and Culture of *Shikoku* Area" had 100 registered students from Kagawa University and 13 students from other universities.

Table 3. eK4 subjects (omnibus style) in 2013.

Subject Provider	Contents Provider (s)	Subject Name	Total
Kagawa Univ.	Tokushima (1), Kagawa (9), Naruto (1), Ehime (2), Shikoku (1)	History and Culture of *Shikoku* Area	113 (13)
Kagawa Univ.	Kagawa (9), Kochi (2), Tokushima Bunri (3)	Local Development of *Shikoku* Area	187 (3)
Kagawa Univ.	Tokushima (1), Kagawa (9), Ehime (1), Kochi (2), Tokushima Bunri (1)	Natural Environment and Disaster Measures of *Shikoku* Area	105 (11)
Total			405 (27)

In many cases, eK4 subjects are provided by single eK4 university. Table **4** shows eight regular style subjects provided in 2013. They are classified into two *Shikoku-gaku* subjects and six interdisciplinary specialized subjects. As for the *Shikoku-gaku* subjects, "*Awa* Studies" treats with *Shikoku* pilgrims called *Ohenro* and includes real walking pilgrims exercise called *aruki-henro*. "Local government theory" considers local autonomy and community development based on histories and future events of *Shikoku*. As for the interdisciplinary specialized subjects, two subjects ("Disaster Prevention of Seismic and Volcanic Hazards" and "Disaster of Floods and Waves") give lectures about disaster prevention. People who lives in *Shikoku* need to have a lot of knowledge about disaster prevention, because natural disasters such as typhoon and earthquake which have often given severe damages to *Shikoku* area. On the hand, in order for students to understand e-Learning appropriately, eK4 provides "Computers in Education" which instructs "what is e-Learning", "how to study with e-Learning", and "where pitfalls are in e-Learning" including the viewpoint of educational technology.

Interdisciplinary specialized subjects generally give special lectures based on fundamental knowledge of *Shikoku*, but some of them like "Computers in Education" treat with peripheral contents in close relations to eK4 activities.

Table 4. eK4 subjects (regular style) in 2013.

Subject Provider	Subject Type	Subject Name	Total
Naruto Univ. of Edu.	*Shikoku-gaku*	*Awa* (*Tokushima*) Studies	75 (0)
Tokushima Bunri Univ.		Local Government Theory	78 (0)
Tokushima Univ.	Interdisciplinary specialized subjects	Guidance of Human Intelligence	19 (0)
Tokushima Univ.		Disaster Prevention of Seismic and Volcanic Hazards	48 (0)
Kagawa Univ.		Computers in Education	118 (10)
Ehime Univ.		Introduction to Future ability	14 (4)
Kochi Univ.		Disaster of Floods and Waves	54 (4)
Tokushima Bunri Univ.		Information Oriented Society Theory	107 (2)
Total			513 (20)

Results of e-Learning Activities in Seven Years

Fig. (**10**) shows the total number of developed e-Learning contents each academic year. Our consortium has started making e-Learning since 2008 when our projects was adopted as a Support project for strategic university collaborations. One content is generally constructed for providing a ninety minute lecture. The total number of developed e-Leaning contents in 2014 reached 482. We construct an e-Learning subject by gathering these contents. Usually, e-Learning consists of fifteen or sixteen e-Learning contents. In order to provide various kinds of e-Learning contents for eK4, some e-Learning contents are made based on concreate theme and the other e-Learning contents are made based on topics with interesting, state of the art and so on. The former contents are top-down approached and the latter contents are bottom-approached. Because an e-Learning subject must have appropriate contents based on a special theme, many developed contents with bottom-up approach are still ready to use now.

Fig. (**11**) shows the number of e-Learning subjects provided for eK4 universities each academic year. We have no data both in 2008 and 2009, because e-Learning with credit transfer system has started since 2010. The number of e-Learning contents in 2011 was decreased, because of re-construction of e-Learning contents. Actually, we found several problems for running e-Learning subjects in eK4. We had one synchronous e-Learning subject in 2010. This e-Learning subject needed support staffs for each universities every time so that it was hard to continue this subject next academic year. In addition, two *Shikoku-gaku* subjects was temporary combined into one subject. The combined subject was provided for only 2011. The academic year of 2011 was the year of trial and error to eK4. The number of e-Learning subjects have gradually increased and it reached fourteen subjects now.

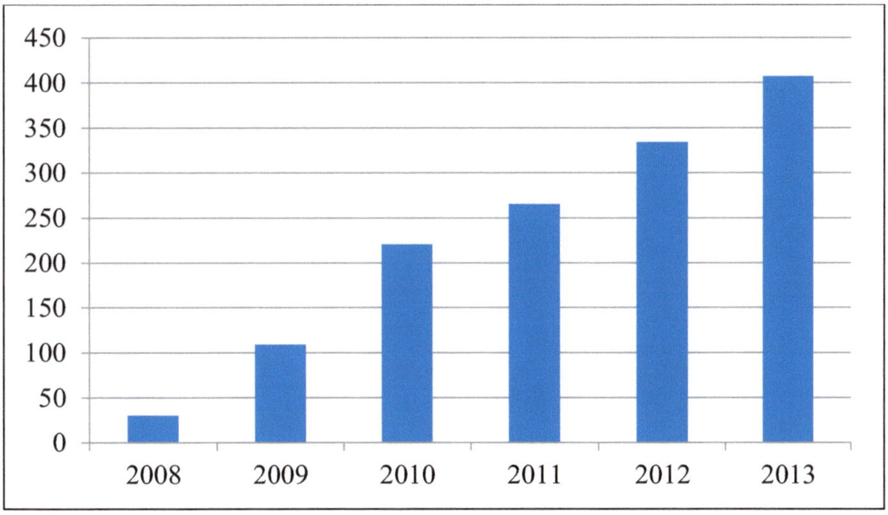

Fig. (10). The number of e-Learning contents.

Fig. (**12**) shows the number of total e-Learning subject registration each academic year. The total number of registration passed over 900 students in 2012 and 2013 but it decreased to 771 students in 2014. We need to analyze the reasons of the declaration, although it is hard to judge whether this is temporally or not. Most of registered students are not special registered students but also students who registered own university subjects. This is a typical feature of eK4 credit transfer system for e-Learning.

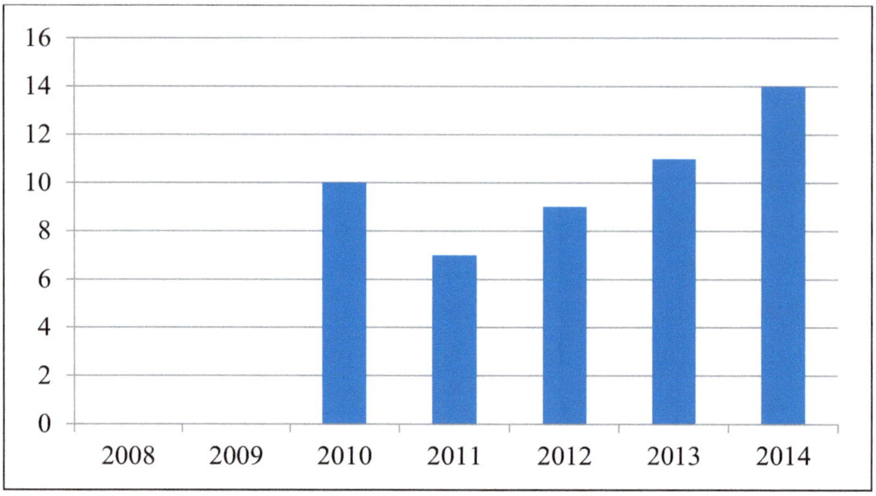

Fig. (11). The number of e-Learning subjects.

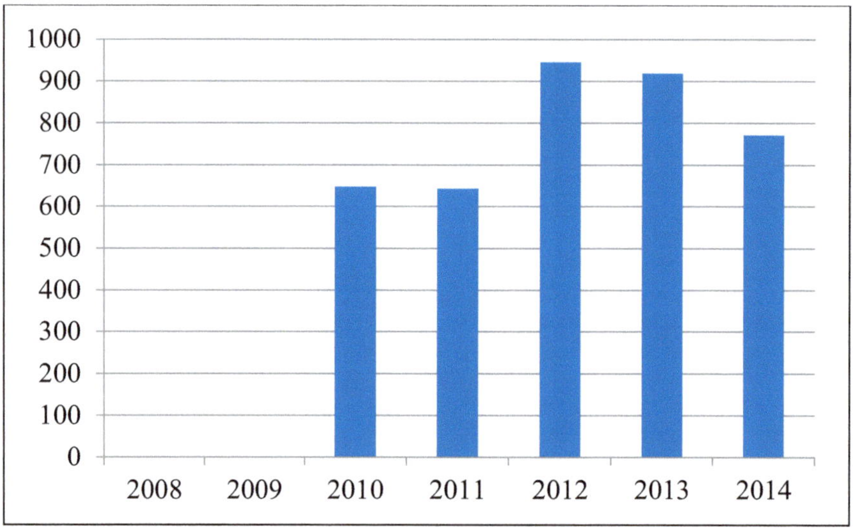

Fig. (12). The number of total e-Learning subject registration.

Issues of Credit Transfer System

As mentioned above, most of registered students are not special registered students but also students who registered own university subjects in e-Learning of eK4. It is an important issue for running credit transfer system in eK4. The followings are revealed reasons for students not to take e-Learning subjects with using credit transfer system.

1. It is not easy for students to find the information e-Learning subjects by credit transfer in syllabus and so on, because these subjects are provided by other universities.
2. Additional application form submission is needed for taking the e-Learning subjects provided other universities compared with own university subjects.
3. It needs much time to transfer credits of the e-Learning subjects provided other universities to own university credit.

The credit transfer system of eK4 is very complicated and inconvenient for students. More or less adjustment does not improve the credit transfer system. The solution to this issue is described in next section.

TO THE NEXT STAGE

Five national university (Tokushima University, Naruto University of Education, Kochi University, Ehime University and Kagawa University) in *Shikoku* have started a new project to establish a knowledge platform through cooperation among them. This project has also been adopted as a project "National University Reform Plan" promoted by MEXT of Japan with five year governmental budget in 2013.

This project has three sub-projects. One of them is e-Learning project called "Cooperative University e-Learning Project for the Shikoku Area". This project was planned based on the experience of e-Learning activities by eK4 project. This project is called "Chipla-e" (*chi* = knowledge in Japanese + platform using e-Learning). The purpose of Chipla-e project is to improve the quality of university education by sharing and complementing human resources and characteristic education and research area of each university, based on mutual cooperation among the five national universities.

For promoting this project, they established "University Consortium for e-Learning, Shikoku Center" (University Consortium for e-Learning, Shikoku Center, n.d.) in the five universities: Kagawa University has the headquarters and the other universities have the branches. This project also uses e-Learning infrastructure of eK4 and improves it. Fig. (**13**) shows conceptual structure of Chipla-e project.

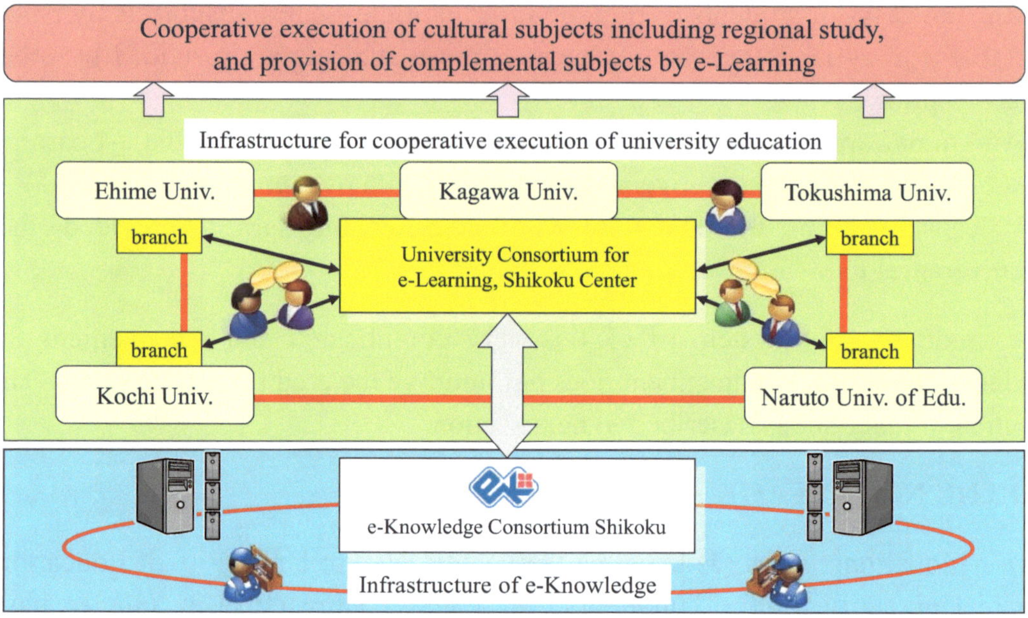

Fig. (13). Structure of Chipla-e project.

Chipla-e project is to realize cooperative education. However, it is not education based on credit transfer system. The main point of cooperative education of Chipla-e is that one professor provides same subjects to five universities. Fig. (**14**) shows the execution model of cooperative education in five *Shikoku* universities. First, a professor of Kagawa University provides an e-Learning subject "X" as the subject of Kagawa University. Next, the professor becomes a part-time instructor of the other universities. Then, the professor provides the subject "X" as the subject of each university. Therefore, cooperative education subjects become own university subjects for students of each university. Cooperative education clears complicated and inconvenient process of credit transfer system of eK4.

Chipla-e project will provide seven e-Learning subjects in the framework of cooperative education in 2015. The number of subjects will increase year by year. In addition, these subjects are also provided to other three universities of eK4 (Shikoku University, Tokushima Bunri University and Kochi University of Technology) by using the credit transfer system of eK4.

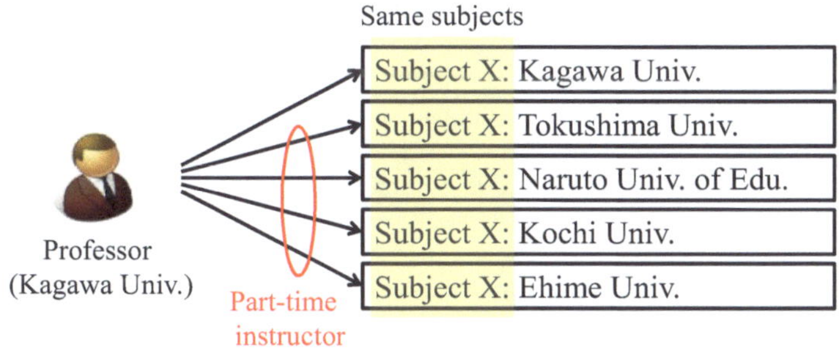

Fig. (14). Execution model of cooperative education in five *Shikoku* universities.

CONCLUSION

This chapter described eK4 project prepared basic infrastructure of e-Leaning environment in *Shikoku* for training talented persons who have high specialty and think of *Shikoku* region. The e-Learning Infrastructure of eK4 has features of distributed LMS and Shibboleth identification. eK4 has started e-Learning based on credit transfer system since 2010. The framework of credit transfer is complicated for many students to take many e-Learning subjects. Currently, Chipla-e project will start cooperative education among five national universities from 2015. Scale up of e-Learning activity is expected by the cooperative education.

CONFLICT OF INTEREST

The author confirms that the author has no conflict of interest to declare for this publication.

ACKNOWLEDGEMENTS

Declared none.

REFERENCES

e-Knowledge Consortium Shikoku. (n.d.). Retrieved August 23, 2015, from http://www.ek4.cc.kagawa-u.a-c.jp/

Institute of Liberal Arts and Sciences (n.d.). Retrieved August 23, 2015, from http://kyoto3univ.jp/

Mistry of Education, Culture, Sports, Science and Technology of Japan. (n.d.). *Support Project for Strategic*

University Collaborations. Retrieved August 23, 2015, from http://www.mext.go.jp/a_menu/koutou/kaikaku/senryaku2.htm

Shibboleth. (n.d.). Retrieved August 23, 2015, from https://shibboleth.net/

The Consortium of National Universities in Hokkaido for Liberal Arts Education. (n.d.). Retrieved August 23, 2015, from http://nucla-hokkaido.jp/

University Consortium for e-Learning, Shikoku Center. (n.d.). Retrieved August 23, 2015, from http://chipla-e.itc.kagawa-u.ac.jp/

University Consortium Saga. (n.d.). Retrieved August 23, 2015, from http://www.saga-cu.jp/

Note: Thes references are all written in Japanese except Shibboleth, *If you have any questions, please feel free to contact the author: hayashi@eng.kagawa-u.ac.jp.*

Telecommunication, 2017, *Vol. 1*, 137-157

CHAPTER 9

Information Technology is Changing People's Transactive Memory

Kuang-Ting Cheng[*]

Department of Information Management, National United University, MiaoLi, Taiwan, R.O.C.

Abstract: Transactive Memory Theory was proposed by Wegner (1986). He considers that in addition to the memory owned by the individual, human has the ability to store memory with the aid of the external reality. Such externally stored memories contain messages that people rely on the memories of one and another other than memorandums and directory data. For instance, John has no idea of Mary's birthday, but he knows Jane remembers it. Therefore, he can know Mary's birthday by asking Jane instead of memorizing it by himself. In the studies for nearly 30 years in the past, the transactive memory theory proposed by Wegner (1986) has almost been applied to the level of the team. However, with the advancement of information technology and popularity of Internet, transactive memory is no longer limited to discussion of team work; rather, it can be further applied to studies in the individual level, just like Google effect and Cloud application.

Keywords: Google effect, Knowledge management, Mental model, Transactive memory systems.

THE DEFINITION OF TRANSACTIVE MEMORY THEORY

The 21st century is an age of innovation. Basically, innovation is founded on multidisciplinary integration, and the basic unit of such integration is knowledge. Therefore, knowledge management is valued by professionals and scholars gradually. Knowledge is the product of an individual with the characteristic of subjectivity. Knowledge of the individual is experience generated from the indivi-

[*] **Corresponding author Kuang Ting Cheng:** Department of Information Management, National United University, MiaoLi, Taiwan, R.O.C.; Tel/Fax: 886-37-381512; E-mail: KTCheng@nuu.edu.tw

Kuo Hung Huang (Ed.)

dual's deduction and prediction of an event's status through his/her mental model. As Craik (1943) defined, mental model is a kind of dynamic characterization and simulation of the external reality. In his viewpoint, the individual tends to convert the external activities into internal models, with which he or she predicts then current conditions. Also, Norman (1983) regarded that the mental model is the internal phenomenal characterization transformed from the individual's continuous interaction with the external. As for Johnson-Laird (1983) and Gilbert and Boulter *et al.* (2000), the former held that the mental model the external phenomenal characterization transformed from the individual's continuous interaction with the internal, and the latter proposed that it is a kind of composite model over the structure of the external reality and a kind of private recognition of the individual.

In the studies of team knowledge, the mental model is a factor proposed for discussions by numerous scholars. Mohammed and Dumville (2001) indicated that the mental model is a kind of knowledge structure owned by the member individual in the team to organizationally aid to describe and explain the state where individual is situated, so that he or she can be guided to interact with other team members in the required environment. To probe into the team's mental model, Klimoski and Mohammed (1994) suggested that it can be divided into three parts: the content refers to knowledge of the team's current condition and the activities they are engaged. Knowledge of the team's current condition consists of the environment it is situated and the issues it has to handle. As for knowledge of the activities the team is engaging, it includes what they are working for and what are its routine tasks. The part of the content can be classified into the environment model with environment related knowledge, the task mental model with knowledge of the team's tasks, the team member mental model with knowledge aiming to understand who knows what among the team members, and finally, the teamwork schema model with knowledge of the task treatment process. The 2nd part of the team's mental model is form that indicates how the team's knowledge is represented organizationally, such as the causal relationship of an event or the way to categorize knowledge systematically. The last one is function that involves three patterns to save the team's knowledge in the team; that is, common knowledge share by the team members or divergent knowledge owned by divergent team members. The third is the mixture pattern, referring to common knowledge and different knowledge owned by the members in the team.

The knowledge of an individual is restricted in the team, it is impossible for merely an individual to store all knowledge required by the whole team. As a result, when the individual encounters the issue unable to be solved solely by his or her knowledge, it is necessary to search for it from other members timely, and the transactive memory is generated whereof. Transactive Memory Theory was proposed by Wegner (1986). He considers that in addition to the memory owned by the individual, human has the ability to store memory with the aid of the external reality. Such externally stored memories contain messages that people rely on the memories of one and another other than memorandums and directory data. For instance, John has no idea of Mary's birthday, but he knows Jane remembers it. Therefore, he can know Mary's birthday by asking Jane instead of memorizing it by him.

Wegner (1995) further illustrated transactive memory, claiming that the team must encode, store, and communicate messages coming from different knowledge territories. So he proposed a three-phase formation and maintenance of the team's transactive memory: Directory Update, Information Allocation, and Retrieval Coordination.

1. Directory Update: The team member is aware of the territories dedicated by the other members in the team, so by the process of the team members' communicating and interacting with one another or carrying out tasks together, a commonly shared foundation is established. With such foundation, he or she can not only enhance his or her devoted territory consistently, but also understand other members' through sharing experiences. With long-term accumulation, they will be able to clearly understand who knows what, and who is good at what.
2. Information Allocation: The team member delivers new information other than what he or she is good at to the expert in that territory in the team, which means to store new information of the team in the most applicable member. In Wegner's point-of-view, if each team member has to memorize all knowledge in the team, it will increase his/her burden of recognition and therefore leads to

dispersed memory and failure of storing the team's memory organizationally. Therefore, allocating proper information to the applicable team members is favorable for saving the knowledge in the team.

3. Retrieval Coordination: When the team member's knowledge is too insufficient to complete the mission, he or she will resort to the members specializing in relative territory, and such process is called retrieval coordination. The team often encounters a lot of diversified tasks; when the team member finds the knowledge he or she possesses is not enough to carry out the task, how to efficiently elicit useful knowledge from other expert members becomes a crucial lesson. In the team, retrieval coordination is founded on the familiarity among the team members, and the degree of each member's understanding of the others' knowledge. If their inter-relationship is tight, knowledge retrieval will be more coordinative.

In 2003, Scholar Lewis integrated and organized the studies on transactive memory in the past, and re-proposed the three dimensions of transactive memory; namely, Specialization, Credibility, and Coordination:

1. Specialization: The structure of knowledge in the team involves whether each team member is equipped with the professional knowledge related to the team's tasks, whether the professional he or she possesses is indispensable for complete the task, and what he or she values is the professional collaboration and supplementation of the structure of knowledge in the team.
2. Credibility: During performing the task, the degree that the team member trust the knowledge and information provided by other members.
3. Coordination: It relates to how team members fully integrate and utilize knowledge and strengths of one and another. Such coordination allows the team members to combine knowledge and strengths of one and another quickly, and depends on whether the team members have clearly understand who knows what and how to make their strengths well fit one and another.

In 2011, Lin unified transactive memory theory and knowledge integration related literature and proposed that transactive memory in the team should be categorized into cognition stage and behavior stage. Cognition stage should contain whether knowledge in the team is with specialization and credibility, as literature in the past pointed out. On the other hand, behavior stage is behavioral integration proposed by studies in the past. Regarding Behavioral Integration, it is composed by Expertise Integration, Collective Mind, and Joint Decision-Making:

1. Expertise Integration: It relates to the quality and quantity of messages inter-exchanged by the team members with one another.
2. Collective Mind: It means the behavior of the team members cooperating with one another.
3. Joint Decision-Making: It refers to the degree that the team members make important decisions together.

MEAUREMENT OF TRANSACTIVE MEMORY

In earlier times, most transactive memory studies adopted laboratory experimentation. The extent of transactive memory was measured through the process of experiment, and the researchers measured the transactive memory system by memory measurement method, behavior observation method, and self-report method. In the light of memory measurement method, by means of measuring the quantity, content, and framework of knowledge and messages recalled with the object of the team, the existence and application condition of transactive memory system was inferred. Most researchers apply comparative analysis of different combinations' learning and memorizing condition of task completion under different environments. Through similar method, Moreland and his coworkers (1999) analyzed personal memory and compared the range and content of collective memory to certify whether transactive memory existed among the team memory. Behavior observation method is a method that conducts observation and evaluation of the team under experimental status. As for Liang (1995), he recorded the process of the experiment with whole course video recording in collaboration with questionnaire results. Then, he performed analysis and judgment to confirm whether transaction memory existed in each team, and implemented the experimental task by transactive memory. In Moreland and his colleagues' (2000) study, they assessed the specialization, coordination, and credibility of the team's memory with 7-point Likert Scale through observing each team's task conducting condition. In regard of self-report method, to reflect transactive memory, interviews, questionnaires, or scales are applied to obtain responses of the tastes after receiving the experiments, such respective knowledge framework, the degree of replying on credibility of knowledge, and the condition of team members' coordinating to tackle knowledge effectively.

The most insufficiency of the three methods mentioned above lies in that the Experiment Group and Control Group have the same tasks and the only and definite solution. Nevertheless, in reality, the method for the team to solve problems is not limited to one, and it is full of uncertainty. In addition, the tasks vary with different items and teams, showing that the external validity of the results of experiment method is obviously insufficient. In recent years, many scholars have started to develop the scales to measure transactive memory. Faraj and Sproull (2000) adopted the concept of expertise coordination to get the location of the team's expertise. Through team management, we can learn the location that demands expertise so as to coordinate and obtain the expertise and rely on knowledge of the team. Yoo (2001) employed studies of Faraj and Sproull to consider the three-question scale of “where is knowledge” and measure the extent of transactive memory in the virtual team. Faraj and Sproull used the foundation of the transactive memory research in the past and developed the three dimensions of transactive memory; that is, Specialization, Credibility, and Coordination. With 124 laboratory teams, 64 item consultant group composed by students with business management expertise, and 27 actual working teams from high technology companies, the research developed the transactive memory scale according to the strict scale development process, and further proved that the scale has better validity and reliability. Kanawattanachai and Yoo (2007) proposed another measurement of three dimensions, including Expertise Location, Cognition-Based Trust, and Task-Knowledge Coordination. Their research results show that people obtain Expertise Location Cognition-Based Trust through the team's transactive memory first, followed by influence on the team's performance with Task-Knowledge Coordination.

STUDIES ON TRANSACTIVE MEMORY THEORY

1. The Influential Factors of Transactive Memory Theory

In generally, transactive memory is generated within the working team, and develops with time. Therefore, the influence of "communication" cannot be ignored. Palazzolo and Serb (2006) found from their research that a good transactive memory system can develop through communication. Jun and Butler (2007) agreed with that, and they discovered the merits and demerits of the transactive memory system with the laboratory experiment method to log the times and ways of the team's communication, like face-to-face communication or electronic media based communication. Hollingshead (1998) held that, in the newly established team, the members normally utilize four communication forms. First, they certify whether their initial concepts towards other members are correct. Next, they more understand other members' professional territory *via* communication, closely followed by achieving a consensus with other members through such communication. Finally, they can definitely clarify respective knowledge territory and take responsibility of it. If there is shortage of communication in the team, information will not be shared and discussed, and messages will not be examined in order to find out the optimal solution. Communication can also be used to keep the teaching obtained from other members in mind to avoid making the same mistake again. Since the quality of transactive memory is affected by communication, consistent memory will be formed among the team members, and messages and experiences will be shared on the foundation of such consistent memory through continuous communication.

In addition to communication, "training" is considered as a factor that affects transactive memory. Brandon and Hollingshead (2004), and Wegner (1995) all regard the formation of the team members' training time and training methods does influence the team's exchangement of memory, among which the training that the team members take part together leads to better performance of exchanging memory. As Brandon and Hollingshead (2004) held, not only training but also "dependency" is one of the factors that influence transactive memory. In this regard, dependency is categorized into "dependency on work" and "dependency on the team". The former refers to the work tasks the team members' face and the degree that they depend on one another. As for the latter, it can be further subdivided into common language, shared experience, and joint decision making. Common language allows the individuals to share and integrate various kinds of knowledge. As the common language is formulated more soundly, integration of knowledge becomes more efficient. When one is unfamiliar with knowledge owned by the other team members, he or she has to integrate various kinds of knowledge through different kinds of contact methods, including language, other forms of symbol communication systems, professional knowledge, shared implication and cognition of individual knowledge territory. In respect with taking part in decision-making together, it means that when the team members face the work tasks, decision-making is not dominated by a single member; rather, it is done through discussion conducted by the team members with one another to make the final decision. The factors that influence transactive memory are listed in Table **1**.

Table 1. Factors that influence transactive memory.

Year	Author	Influential Factors of Transactive Memory
1998	Hollingshead, Andrea B.	The research found that face-to-face communication can make transactive memory better. (Andrea B Hollingshead, 1998)
1991	Wegner, Daniel M. Erber, Ralph Raymond, Paula	The research found the close relationship among the team members influences transactive memory. (Wegner, Erber, & Raymond, 1991)
2000	Hollingshead, Andrea B.	*Via* experiments, it was found that the common experiences and close relationship of the team members influences transactive memory in the team. (Andrea B Hollingshead, 2000)
2000	Moreland, Richard L. Myaskovsky, Larissa	The research found that through group training, transactive memory will be better. (Richard L. Moreland & Myaskovsky, 2000)
2000	Rulke, Diane Liang Rau, Devaki	The experiment found that training and communication influence transactive memory. (Rulke & Rau, 2000)
2001	Mohammed, Susan Dumville, Brad C.	The research held that team model includes transactive memory, which is influenced by training. (Mohammed & Dumville, 2001)
2003	Lewis, Kyle	The research developed measurement scale for transactive memory, and measure the three dimensions, specialization, credibility, and coordination, transactive memory was measured. (Lewis, 2003)
2003	Hollingshead, Andrea B. Brandon, David P.	The research considers that transactive memory must be recycled by the team continuously, including proper work, experiences, and members. (Andrea B. Hollingshead & Brandon, 2003)
2004	Brandon, David P. Hollingshead, Andrea B.	Based on the research results, it is considered that transactive memory will be better with interaction among the task, experience, and members. (Brandon & Hollingshead, 2004)
2004	Fraidin, Samuel N.	He indicated that the number of team members influences transactive memory in the team. (Fraidin, 2004)

(Table 1) contd.....

Year	Author	Influential Factors of Transactive Memory
2004	Lewis, Kyle	Through the finding of transactive memory theory, it was found that knowledge influences performance. Besides, for knowledge working team, communication methods and frequency will influence the quality of transactive memory. (Lewis, 2004)
2005	Akgün, Ali E. Byrne, John Keskin, Halit Lynn, Gary S. Imamoglu, Salih Z.	The team's stability, credibility, amiability, familiarity, and communication were discussed by questionnaires Those factors will influence transactive memory in the team, and further influence generation of the project. Among them, complexity of the task is the mediate variable. (Ali E Akgün, Byrne, Keskin, Lynn, & Imamoglu, 2005)
2005	Lewis, Kyle Lange, Donald Gillis, Lynette	Ceaseless learning and practice influences transactive memory. (Lewis, Lange, & Gillis, 2005)
2005	Nevoa, Dorit Wand, Yair	By means of transactive memory, it discussed the memory in the organization. The research claimed that knowledge in small groups is very important. (Nevoa & Wand, 2005)
2005	Reagans, Ray Argote, Linda Brooks, Daria	From findings in the experiments by the teaching hospital, it was found that increase of personal experience and knowledge influences transactive memory. (Reagans, Argote, & Brooks, 2005)
2006	Pearsall, Matthew J. Ellis, Aleksander P.J.	The research found that strong pressure influences negatively on the mental model and transactive memory. (Pearsall & Ellis, 2006)
2006	Rau, Devaki	Transactive memory increases with time, and information acquiring ability and information correctness improves as will. (Rau, 2006)
2006	Palazzolo, Edward T. Serb, Dana A. She, Yuechuan Su, Chunke Contractor, Noshir S.	The organization can elicit knowledge in transactive memory. Communication is an important factor. (Palazzolo *et al.*, 2006)
2006	Yuqing, Ren Carley, Kathleen M. Argote, Linda	The research explored that transactive memory's effect in different groups and groups with different sizes. (Yuqing, Carley, & Argote, 2006)
2007	Lewis, Kyle Belliveau, Maura Herndon, Benjamin Keller, Joshua	By means of experiment method, discussion on the structure and process of transactive memory was conducted. The research found that the key to team member update and the team's working efficiency lies in whether the experienced members aggregate knowledge on work. (Lewis, Belliveau, Herndon, & Keller, 2007)
2007	Majchrzak, Ann Jarvenpaa, Sirkka L. Hollingshead, Andrea B.	By using transactive memory, it discussed how the team established in emergency coordinates knowledge, and extends the three constructs proposed by Lewis. (Majchrzak, Jarvenpaa, & Hollingshead, 2007)
2007	Prichard, Jane S. Ashleigh, Melanie J.	The research team's skill training influences transactive memory. (Prichard & Ashleigh, 2007)
2007	Sharma, Rajeev Yetton, Philip	Training influences the success rate of conducting a task. Transactive memory can function as the regulating variable. (Sharma & Yetton, 2007)

(Table 1) contd.....

Year	Author	Influential Factors of Transactive Memory
2007	Yuan, Y.Connie Fulk, Janet Monge, Peter R.	By the ad hoc experimental observation of the team in operation, how the transactive memory system is formed was learned. (Yuan, Fulk, & Monge, 2007)
2008	Jackson, Paul Klobas, Jane	With case study, ecology was found in transactive memory theory in different levels of information in the organization. (Jackson & Klobas, 2008)
2009	Liu, Lei Hinds, Pamela J.	The group relationship of human-machine interface is discussed by the transactive memory theory. (L. Liu & Hinds, 2009)
2010	Choi, Sue Young Lee, Heeseok Yoo, Youngjin	Information Technology influences the transactive memory system. (Choi, Lee, & Yoo, 2010)
2012	Zheng, Yanfeng	The common experiences of the team members in the past influences the transactive memory system and further influences efficacy, Similarity of tasks and credibility is important regulating variables. (Zheng, 2012)
2013	Noroozi, Omid Biemans, Harm J. A. Weinberger, Armin Mulder, Martin Chizari, Mohammad	The transactive memory system influence the individual in the team. (Noroozi, Biemans, Weinberger, Mulder, & Chizari, 2013)

2. Application of the Transactive Memory Theory

From organizing literature related to transactive memory theory, it is found that application of the transactive memory theory can be divided into two types. One is studies applied to team performance. Akgün & Lynn (2002) pioneered to use transactive memory theory to explore the stability of the team by taking the team of new product of development. Akgün *et al.* (2005) probed into the team's stability, credibility, amiability, familiarity, and communication by questionnaires to see whether those factors have an influence on the team's transactive memory, and further impacts on the output of the project. The research found that complexity of the task can be regarded as the regulating variable. In the following year, Akgün *et al.* re-applied the transactive memory to the new product developing team. Austin (2000) found from his research that the team's transactive memory and the density of social network has an impact on efficacy. Austin (2003) discovered from continuous studies that, in the organization, transactive memory theory may cast their influence to the groups efficacy through consistency, specialization, and correctness. As for Child & Shumate (2007), they indicated that the transactive memory theory can foster the team's efficiency of the team remarkably, while Kanawattanachai & Yoo (2007) pointed out that communication on working tasks will affect the team's transactive memory and further impact on their efficacy. Cruz *et al.* (2007) found that transactive memory will have an influence on the team's output, and Rau (2006) claimed that transactive memory indeed impact the team's performance. Sharma & Yetton (2007) considered that training affects the success rate of the task, where transactive memory may be used as the regulating variable. Young & Kanawattanachai (2001) used the transactive memory theory to discuss team performance, which found that the collective mind can be considered as the mediate variable. Last of all, Yuqing *et al.* (2006) canvassed the effect of the transactive theory on different groups and groups with different sizes. In the meantime, Zhang (2007) adopted the transactive memory theory to coordinate the characteristics of each team member to further promote their team performance.

The other type of transactive memory application refers to studies on knowledge management. Alavi & Tiwana (2002) held that knowledge management system can be developed through the transactive memory theory and communication. On the other hand, through the transactive memory theory, Borgatti & Cross (2003) explored information searching in organization learning and learning in social network. In Griffith & Neale's (2003) perspective, potential knowledge in the team can be transformed into useful knowledge by means of the individual's absorbing ability and transactive memory in the team. Similarly, Hollingshead (2000) searched for special knowledge with the transactive memory theory, while Jackson & Klobas (2008) discovered the ecology extracted from information in different tiers in the organization *via* the transactive memory theory. As for Majchrzak (2007), the scholars utilized the transactive memory to discuss how the team established in emergency coordinate knowledge by means of the transactive memory theory, as well as extending the three conducts proposed by Lewis. Meanwhile, Nevoa & Wand (2005) used the transactive memory theory to probed into the organization's memory. It is claimed that knowledge exists in the small groups is remarkably important. Rulke & Rau (2000) observed the video tapes and found transactive memory though the group trainings. They thus argued that the individual can develop his or her personal expertise. Application of transactive memory is described in Table **2**.

Table 2. Application of the transactive memory theory.

Year	Author	Application of the Transactive Memory Theory
2000	Austin, John R.	The team's transactive memory and density of social network can influence the efficacy. (John R Austin, 2000)
2001	Yoo, Youngjin Kanawattanachai, Prasert	The collective mind can be considered as the mediate variable when the researchers discussed the team's performance of the transactive memory theory. (Yoo & Kanawattanachai, 2001)
2002	Akgün, Ali E. Lynn, Gary S.	They probed into the team's stability by the transactive memory theory, and took the new product developing team as the research subject. (Ali E. Akgün & Lynn, 2002)
2002	Alavi, Maryann Tiwana, Amrit	Knowledge management system can be developed through the transactive memory theory and communication. (Alavi & Tiwana, 2002)
2003	Austin, John R.	The transactive memory theory may cast their influence to the group efficacy through consistency, specialization, and correctness. (John R. Austin, 2003)
2003	Borgatti, Stephen P. Cross, Rob	By the transactive memory theory, the study explored information searching in organization learning and learning in social network. (Borgatti & Cross, 2003)
2003	Griffith, Terri L. Sawyer, John E. Neale, Margaret A.	Potential knowledge in the team can be transformed into useful knowledge by means of the individual's absorbing ability and transactive memory in the team. (Griffith *et al.*, 2003)
2005	Rau, Devaki	Transactive memory influence on performance. (Rau, 2005, 2006)
2006	Akgün, Ali E. Byrne, John C. Keskin, Halit Lynn, Gary S.	The transactive memory theory is applied to the new product developing team. (Ali E Akgün, Byrne, Keskin, & Lynn, 2006)
2006	Pearsall, Matthew J. Ellis, Aleksander P. J.	Characteristics of team members influence transactive memory and further influence the team's performance. (Pearsall & Ellis, 2006)
2007	Child, Jeffrey T. Shumate, Michelle	The transactive memory theory can foster the team's efficiency of the team remarkably. (Child & Shumate, 2007)
2007	Kanawattanachai, Prasert Yoo, Youngjin	Communication on working tasks will affect the team's transactive memory and further impact on their efficacy. (Kanawattanachai & Yoo, 2007)
2007	Cruz, Natalia Martin Perez, Victor Martin Ramos, Yolanda Fernandez	Transactive memory will have an influence on the team's output. (Cruz *et al.*, 2007)
2007	Zhang, Zhi-Xue Hempel, Paul S. Han, Yu-Lan Tjosvold, Dean	By means of the transactive memory theory, characteristics of each member are coordinated to enhance the team's performance. (Zhang *et al.*, 2007)

(Table 2) contd.....

Year	Author	Application of the Transactive Memory Theory
2010	Gino, Francesca Argote, Linda Miron-Spektor, Ella Todorova, Gergana	The transactive memory system influences the team's creativity. (Gino, Argote, Miron-Spektor, & Todorova, 2010)
2010	Schreiber, Mareike Engelmann, Tanja	Technology tools are helpful for transactive memory, and the transactive memory system influences the team's efficacy. (Schreiber & Engelmann, 2010)
2010	Liu, Bangcheng Zang, Zhi	The transactive memory system influences the efficiency and efficacy. (B. Liu & Zang, 2010)
2012	Hsu, Jack Shih-Chieh Shih, Sheng-Pao Chiang, Jerry C. Liu, Julie Yu-Chih	The transactive memory system influences the efficacy by coordination and cooperation. (Hsu, Shih, Chiang, & Liu, 2012)
2012	Comu, Semra Iorio, Josh Taylor, John E. Dossick, Carrie Sturts	Transactive memory influences the virtual teams around the world. (Comu, Iorio, Taylor, & Dossick, 2012)

APPLICATION OF INFORMATION TECHNOLOGY TO THE TRANSACTIVE MEMORY THEORY

In the studies for nearly 30 years in the past, the transactive memory theory proposed by Wegner (1986) has almost been applied to the level of the team. However, with the advancement of information technology and popularity of Internet, transactive memory is no longer limited to discussion of team work; rather, it can be further applied to studies in the individual level.

Sparrow *et al.* (2011) utilized the laboratory experimental method to conduct their study. They took the students as the research subjects, dividing them into two groups and asking them to record the trifles in daily life. One group was demanded to delete what they had recorded right away, while the other one could save the trifles in file folders in the computer. According to the research results, those who were required to delete all trifle records still remembered what they had written down, but the other group who stored the records in computer could not remember them except the folder that accommodated their files. The scholars held that Internet has changed the ability of human brain to deal with information by making Internet as the memory bank of the individual. That is to say, they have made the computer or Internet as man's external memory, calling such phenomena as “Google effect”.

Research on Google effect tries to treat change of human memory from the psychological viewpoint and the research results based on man's existing memory. It is found that the facilities and equipment of information technology have resulted in significant degeneration of their memory and ability to illustrate things. As transactive memory theory proposed, people only need to know where the illustration of things is saved and get the memory from the external device. It is similar to that people take notes on 3M post-it note at work rather than memorize it in the brain. When they have to remember the content, they can go for the post-it note directly.

In the past, when information technology has not entered the individual territory, people used knowledge in their own memory to solve the issues in life or work. When personal knowledge was not sufficient enough to solve the problem, they usually took two kinds of approaches. One is to enrich knowledge to their memory by books or documents, and the other is to consult their neighbors or coworkers. From the angle of psychology, the first approach focuses on expanding the individual's memory for more knowledge, and the second aims at expanding the external memory by storing knowledge in others' brain. As long as the external memory has the index of such external memory, the individual can therefore be guided to obtain what he or she requires.

With advancement of technology, computer and Internet has entered modern people's life and work, and changed the above mentioned two approaches used to seek for solutions in the past. The birth of “Cloud Technology” permits us to record the desired texts, images, and videos and store them on the external memory called “Cloud”. Such phenomena is like expanding the individual's memory, the only difference is this is an external memory. The only thing the individual memory owned by people has to do is to record the location of the Cloud system that stores data. Whenever knowledge is required, it can be accessed *via* Cloud. YouTube, Instagram, and Dropbox that are popular nowadays are the best examples.

Information technology has changed the second knowledge acquire approach for the individual to solve the problem. Actually, it relies on "Web 2.0" and the "Search Engine". "Web 2.0" means the distributed network phenomena that the user creates, collaborates, and shares various kinds of information (Cheng, Hsu, & Lin, 2012), which is a new concept emerging since 2004. This new concept breaks through the operation model that e-business is controlled collectively by the vendors. The website content is provided by the user voluntarily in order to stimulate more users to visit the website. Currently, the representative of Web 2.0 is Wikipedia, which is based on Wiki tools. On Wikipedia, thousands of users contribute their own knowledge on basis of the Web as the platform, allowing the range of knowledge collected by Wikipedia to expand continuously and limitlessly in a short-term. Knowledge involving astrology, geography, calendar, politics, traffic, human affairs, art, museum, literature, and Confucianism are all included in Wikipedia with more than 253 languages. Such tremendous knowledge database may force the conventional Encyclopedia Britannica to retreat from the market (Cheng *et al.*, 2012). In Taiwan, the well-known Web 2.0 websites are all based on Blog tool. The user can take advantage of the tools provided by the Blog to create and share knowledge. Besides, they can interact with other users on this Web platform in order to accumulate knowledge and brainstorm. When Web 2.0 creates a lot of knowledge create by numerous individuals, a "Search Engine" is then demanded to organize knowledge, and offer a good search mechanism and interface for people to retrieve other people's memory. Google, Yahoo, and Bing are the best examples.

In respect with the transactive memory theory, from people-to-people interaction in the early stage, to people-to-information technology interaction today, it is always concentrated on man's ability to distribute knowledge to different memory devices with various kinds of methods. Before information technology starts to influence mankind, the required knowledge is stored in other individuals that their interpersonal relationship can reach. Now, however, with the emergence of "Cloud Technology", "Search Engine" and "Web2.0", they have turned to store knowledge within the information system. If human beings do not have to memorize so much knowledge, it is supposed that they may endeavor to what is more creative (Sparrow *et al.*, 2011).

CONFLICT OF INTEREST

The author confirms that the author has no conflict of interest to declare for this publication.

ACKNOWLEDGEMENTS

Declared none.

REFERENCES

Akgün, A.E., Byrne, J., Keskin, H., Lynn, G.S., Imamoglu, S.Z. (2005). Knowledge networks in new product development projects: A transactive memory perspective. *Information and Management, 42*(8), 1105-1120. [http://dx.doi.org/10.1016/j.im.2005.01.001]

Akgün, A.E., Byrne, J.C., Keskin, H., Lynn, G.S. (2006). Transactive memory system in new product development teams. *IEEE Transactions on Engineering Management, 53*(1), 95-111. [http://dx.doi.org/10.1109/TEM.2005.857570]

Akgün, A.E., Lynn, G.S. (2002). Antecedents and consequences of team stability on new product development performance. *Journal of Engineering and Technology Management, 19*(3–4), 263-286. [http://dx.doi.org/10.1016/S0923-4748 (02)00021-8]

Alavi, M., Tiwana, A. (2002). Knowledge integration in virtual teams: The potential role of kms. *Journal of the American Society for Information Science Technology, 53*(12), 1029-1037. [http://dx.doi.org/10.1002/asi.10107]

Austin, J.R. 2000. *Knowing what and whom other people know: Linking transactive memory with external connections in organizational groups*. Paper Presented at the 16th Annual Meeting of Academy of Management Meeting. Toronto [http://dx.doi.org/10.5465/APBPP.2000.5535165]

Austin, J.R. (2003). Transactive memory in organizational groups: the effects of content, consensus, specialization, and accuracy on group performance. *Journal of Applied Psychology, 88*(5), 866-878. [http://dx.doi.org/10.1037/0021-9010.88.5.866] [PMID: 14516250]

Borgatti, S.P., Cross, R. (2003). A relational view of information seeking and learning in social networks. *Management Science, 49*(4), 432-445. [http://dx.doi.org/10.1287/mnsc.49.4.432.14428]

Brandon, D.P., Hollingshead, A.B. (2004). Transactive memory systems in organizations: Matching tasks, expertise, and people. *Organization Science, 15*(6), 633-644. [http://dx.doi.org/10.1287/orsc.1040.0069]

Cheng, K-T., Hsu, J.S, Lin, T-C. (2012). Understanding the intention to continue using web 2.0 to transfer and share knowledge. *Journal of International Management, 19*(2), 249-274.

Child, J.T., Shumate, M. (2007). The impact of communal knowledge repositories and people-based knowledge management on perceptions of team effectiveness. *Management Communication Quarterly, 21*(1), 29-54.
[http://dx.doi.org/10.1177/0893318907301987]

Choi, S.Y., Lee, H., Yoo, Y. (2010). The impact of information technology and transactive memory systems on knowledge sharing, application, and team performance: A field study. *Management Information Systems Quarterly, 34*(4), 855-870.

Comu, S., Iorio, J., Taylor, J.E., Dossick, C.S. (2012). Quantifying the impact of facilitation on transactive memory system formation in global virtual project networks. *Journal of Construction Engineering and Management, 139*(3), 294-303.
[http://dx.doi.org/10.1061/ (ASCE)CO.1943-7862.0000610]

Craik, K. (1943). *The nature of explanation.*. England: Cambridge University Press.

Cruz, N.M., Perez, V.M., Ramos, Y.F. (2007). Transactive memory processes that lead to better team results. *Team Performance Management, 13*(7/8), 192-205.
[http://dx.doi.org/10.1108/13527590710842529]

Faraj, S., Sproull, L. (2000). Coordinating expertise in software development teams. *Management Science, 46*(12), 1554-1568.
[http://dx.doi.org/10.1287/mnsc.46.12.1554.12072]

Fraidin, S.N. (2004). When is one head better than two? Interdependent information in group decision making. *Organizational Behavior and Human Decision Processes, 93*(2), 102-113.
[http://dx.doi.org/10.1016/j.obhdp.2003.12.003]

Gilbert, J.K., Boulter, C.J. (2000). *Developing models in science education.* Kluwer Academic Publishers.
[http://dx.doi.org/10.1007/978-94-010-0876-1]

Gino, F., Argote, L., Miron-Spektor, E., Todorova, G. (2010). First, get your feet wet: The effects of learning from direct and indirect experience on team creativity. *Organizational Behavior and Human Decision Processes, 111*(2), 102-115.
[http://dx.doi.org/10.1016/j.obhdp.2009.11.002]

Griffith, T.L., Sawyer, J.E., Neale, M.A. (2003). Virtualness and knowledge in teams: Managing the love triangle of organizations, individuals, and information technology. *Management Information Systems Quarterly, 27*(2), 265-287.

Hollingshead, A.B. (1998). Retrieval processes in transactive memory systems. *Journal of Personality and Social Psychology, 74*(3), 659-671.
[http://dx.doi.org/10.1037/0022-3514.74.3.659]

Hollingshead, A.B. (2000). Perceptions of expertise and transactive memory in work relationships. *Group Processes and Intergroup Relations, 3*(3), 257-267.

[http://dx.doi.org/10.1177/1368430200033002]

Hollingshead, A.B., Brandon, D.P. (2003). Communication and transactive memory systems. *Human Communication Research, 29*(4), 607-615.
[http://dx.doi.org/10.1111/j.1468-2958.2003.tb00859.x]

Hsu, J.S, Shih, S-P., Chiang, J.C., Liu, J.Y (2012). The impact of transactive memory systems on is development teams' coordination, communication, and performance. *International Journal of Project Management, 30*(3), 329-340.
[http://dx.doi.org/10.1016/j.ijproman.2011.08.003]

Jackson, P., Klobas, J. (2008). Building knowledge in projects – a practical application of social constructivism to information systems development. *International Journal of Project Management, 26*(4), 329-337.
[http://dx.doi.org/10.1016/j.ijproman.2007.05.011]

Johnson-Laird, P.N. (1983). *Mental models: Toward a cognitive science of language.* Cambridge: Harvard University Press.

Jun, H.E., Butler, B.S., King, W.R. (2007). Team cognition: Development and evolution in software project teams. *Journal of Management Information Systems, 24*(2), 261-292.
[http://dx.doi.org/10.2753/MIS0742-1222240210]

Kanawattanachai, P., Yoo, Y. (2007). The impact of knowledge coordination on virtual team performance over time. *Management Information Systems Quarterly, 31*(4), 783-808.

Klimoski, R., Mohammed, S. (1994). Team mental model: Construct or metaphor? *Journal of Management, 20*(2), 403.
[http://dx.doi.org/10.1177/014920639402000206]

Lewis, K. (2003). Measuring transactive memory systems in the field: scale development and validation. *Journal of Applied Psychology, 88*(4), 587-604.
[http://dx.doi.org/10.1037/0021-9010.88.4.587] [PMID: 12940401]

Lewis, K. (2004). Knowledge and performance in knowledge-worker teams: A longitudinal study of transactive memory systems. *Management Science, 50*(11), 1519-1533.
[http://dx.doi.org/10.1287/mnsc.1040.0257]

Lewis, K., Belliveau, M., Herndon, B., Keller, J. (2007). Group cognition, membership change,and performance: Investigating the benefits and detriments of collective knowledge. *Organizational Behavior and Human Decision Processes, 103*(2), 159-178.
[http://dx.doi.org/10.1016/j.obhdp.2007.01.005]

Lewis, K., Lange, D., Gillis, L. (2005). Transactive memory systems, learning, and learning transfer. *Organization Science, 16*(6), 581-598.
[http://dx.doi.org/10.1287/orsc.1050.0143]

Liang, D.W., Moreland, R., Argote, L. (1995). Group *versus* individual training and group performance: The mediating role of transactive memory. *Personality and Social Psychology Bulletin, 21*(4), 384-393. [http://dx.doi.org/10.1177/0146167295214009]

Liu, B., Zang, Z. (2010). The mediating effects of team efficacy on the relationship between a transactive memory system and team performance. *Social Behavior and Personality: an International Journal, 38*(7), 865-869.

Liu, L., Hinds, P.J. (2009). *Transactive memory systems: A perspective on coordination in human-robot incident response teams.* Paper presented at the Human-Robot Interaction (HRI), 2009 4th ACM/IEEE International Conference on, La Jolla, CA. [http://dx.doi.org/10.1145/1514095.1514170]

Majchrzak, A., Jarvenpaa, S.L., Hollingshead, A.B. (2007). Coordinating expertise among emergent groups responding to disasters. *Organization Science, 18*(1), 147-161. [http://dx.doi.org/10.1287/orsc.1060.0228]

Mohammed, S., Dumville, B.C. (2001). Team mental models in a team knowledge framework: Expanding theory and measurement across disciplinary boundaries. *Journal of Organizational Behavior, 22*(2), 89-106. [http://dx.doi.org/10.1002/job.86]

Moreland, R.L. (1999). Transactive memory: Learning who knows what in word groups and organizations. In: Thompson, L.L., Levine, J.M., Messick, D.M., (Eds.), *Shared cognition in organizations: The management of knowledge.* (pp. 3-31). NJ: Lawrence Erlbaum Associates.

Moreland, R.L., Myaskovsky, L. (2000). Exploring the performance benefits of group training: Transactive memory or improved communication? *Organizational Behavior and Human Decision Processes, 82*(1), 117-133. [http://dx.doi.org/10.1006/obhd.2000.2891]

Nevoa, D., Wand, Y. (2005). Organizational memory information systems: A transactive memory approach. *Decision Support Systems, 39*(4), 549-562. [http://dx.doi.org/10.1016/j.dss.2004.03.002]

Norman, D.A. (1983). Some observations on mental models. In: Gentner, D., Stevens, A.L., (Eds.), *Mental Models* Lawrence Erlbaum Associates.

Noroozi, O., Biemans, H.J., Weinberger, A., Mulder, M., Chizari, M. (2013). Scripting for construction of a transactive memory system in multidisciplinary cscl environments. *Learning and Instruction, 25*(0), 1-12. [http://dx.doi.org/10.1016/j.learninstruc.2012.10.002]

Palazzolo, E.T., Serb, D.A., She, Y., Su, C., Contractor, N.S. (2006). Coevolution of communication and knowledge networks in transactive memory systems: Using computational models for theoretical development. *Communication Theory, 16*(2), 223-250. [http://dx.doi.org/10.1111/j.1468-2885.2006.00269.x]

Pearsall, M.J., Ellis, A.P. (2006). The effects of critical team member assertiveness on team performance and satisfaction. *Journal of Management, 32*(4), 575-594.
[http://dx.doi.org/10.1177/0149206306289099]

Peltokorpi, V. (2004). Transactive memory directories in small work units. *Personnel Review, 33*(4), 446-467.
[http://dx.doi.org/10.1108/00483480410539515]

Prichard, J.S., Ashleigh, M.J. (2007). The effects of team-skills training on transactive memory and performance. *Small Group Research, 38*(6), 696-726.
[http://dx.doi.org/10.1177/1046496407304923]

Rau, D. (2005). The influence of relationship conflict and trust on the transactive memory: Performance relation in top management teams. *Small Group Research, 36*(6), 746-771.

Rau, D. (2006). Top management team transactive memory, information gathering, and perceptual accuracy. *Journal of Business Research, 59*(4), 416-424.
[http://dx.doi.org/10.1016/j.jbusres.2005.07.001]

Reagans, R., Argote, L., Brooks, D. (2005). Individual experience and experience working together: Predicting learning rates from knowing who knows what and knowing how to work together. *Management Science, 51*(6), 869-881.
[http://dx.doi.org/10.1287/mnsc.1050.0366]

Rulke, D.L., Rau, D. (2000). Investigating the encoding process of transactive memory development in group training. *Group and Organization Management, 25*(4), 373-396.
[http://dx.doi.org/10.1177/1059601100254004]

Schreiber, M., Engelmann, T. (2010). Knowledge and information awareness for initiating transactive memory system processes of computer-supported collaborating ad hoc groups. *Computers in Human Behavior, 26*(6), 1701-1709.
[http://dx.doi.org/10.1016/j.chb.2010.06.019]

Sharma, R., Yetton, P. (2007). The contingent effects of training, technical complexity, and task interdependence on successful information systems implementation. *Management Information Systems Quarterly, 31*(2), 219-238.

Sparrow, B., Liu, J., Wegner, D.M. (2011). Google effects on memory: cognitive consequences of having information at our fingertips. *Science, 333*(6043), 776-778.
[http://dx.doi.org/10.1126/science.1207745] [PMID: 21764755]

Wegner, D.M. (1986). Transactive memory: A contemporary analysis of the group mind. In: Mullen, B., Goethals, G.R., (Eds.), *Theories of group behavior.* (pp. 185-205). New York: Springer-Verlag.

Wegner, D.M. (1995). A computer network model of human transactive memory. *Social Cognition, 13*(3), 319-339.

[http://dx.doi.org/10.1521/soco.1995.13.3.319]

Wegner, D.M., Erber, R., Raymond, P. (1991). Transactive memory in close relationships. *Journal of Personality and Social Psychology, 61*(6), 923-929.
[http://dx.doi.org/10.1037/0022-3514.61.6.923] [PMID: 1774630]

Yoo, Y., Kanawattanachai, P. (2001). Developments of transactive memory systems and collective mind in virtual teams. *International Journal of Organizational Analysis, 9*(2), 187-208.
[http://dx.doi.org/10.1108/eb028933]

Yuan, Y.C., Fulk, J., Monge, P.R. (2007). Access to information in connective and communal transactive memory systems. *Communication Research, 34*(2), 131-155.
[http://dx.doi.org/10.1177/0093650206298067]

Yuqing, R., Carley, K.M., Argote, L. (2006). The contingent effects of transactive memory: When is it more beneficial to know what others know? *Management Science, 52*(5), 671-682.
[http://dx.doi.org/10.1287/mnsc.1050.0496]

Zhang, Z-X., Hempel, P.S., Han, Y-L., Tjosvold, D. (2007). Transactive memory system links work team characteristics and performance. *Journal of Applied Psychology, 92*(6), 1722-1730.
[http://dx.doi.org/10.1037/0021-9010.92.6.1722] [PMID: 18020808]

Zheng, Y. (2012). Unlocking founding team prior shared experience: A transactive memory system perspective. *Journal of Business Venturing, 27*(5), 577-591.
[http://dx.doi.org/10.1016/j.jbusvent.2011.11.001]

CHAPTER 10

Using Picture Book Apps to Assist Children's Heritage Language Self-Learing

Yu-Ching Yeh[1], **Hsiang-Ju Ho**[1,*] and **Ming-Chung Chen**[2]

[1] *Department of Early Childhood Education, National Chiayi University, Chiayi, Taiwan*

[2] *Department of Special Education, National Chiayi University, Chiayi, Taiwan*

Abstract: Prior studies have shown that information technology is effective in increasing children's vocabulary learning and reading comprehension; however, its effect on bilingual or heritage language learning needs further investigation. This study examines the effect of a self-learning vocabulary program by using bilingual picture book apps designed employing the universal design for learning approach. Data were collected from 39 primary school pupils from international marriage families in Taiwan. The pupils were randomly assigned to experimental and control groups. The experimental group used the tablets to read picture book apps while the control group used printed picture books. The results showed that the experimental group performed better on a Vietnamese vocabulary test than the control group. The results also indicated that the pupils with higher proficiencies in their first language performed better in Vietnamese vocabulary learning. Tablet picture book apps can be effective for supporting bilingual learning because they are easy to operate and more fun to use than traditional books. In addition, tablet picture books consist of a story context, word games, and sounds, thus motivating children to learn the heritage language actively.

Keywords: Bilingual, Computer-assisted language learning, Cross-national marriage, Heritage language, Information communication technology (ICT), Mobile learning, Picture book apps, Self-learning, Tablet, Universal design for learning (UDL).

* **Corresponding author Hsiang-Ju Ho:** Department of Early Childhood Education, National Chiayi University, Chiayi, Taiwan; Tel/Fax: +886-5-2269304/+886-5-2260582; E-mail: hjho@mail.ncyu.edu.tw

Kuo Hung Huang (Ed.)

INTRODUCTION

Immigrant children use their heritage language less the longer they stay in the host society. Maintenance of the heritage language depends on the immigrant parents' positive attitude (Park & Sarkar, 2008; Yeh, 2012). Tseng and Fuligni (2000) found that immigrant parents tend to shift from their heritage language to English when they see that their adolescent children prefer to speak English at home. Park, Tsai, Liu, and Lau (2012) also suggested that immigrant parents tend to modify their own language behaviour according to their children's language use and proficiency. Immigrant parents may expect their children to master the host language instead of the heritage language, for the sake of their children's future development (Uttal & Han, 2011). However, children learn their heritage language with great difficulty when their family cannot provide a supportive environment.

Although heritage language acquisition and maintenance builds children's confidence and enables their development (Nomura & Caidi, 2013), it is not given adequate attention in education. Heritage language is seldom included as part of the school curriculum (Moin, Schwartz, & Breitkopf, 2011), and it rarely achieves a status equivalent to that of a national standard language (Spolsky, 2011).

A heritage speaker acquires the heritage language as his/her first language through natural input in the home environment (Montrul, 2002). Since the heritage language is transmitted in the form of 'daily conversations' with the family, language learning materials or pedagogy may not be taken into consideration. However, it is difficult to expect that families alone should teach children given the limitations of immigrant parents' capabilities. The barriers include parents' education level, improper instruction methods and materials, and time limitations (National Immigration Agency 2010; Valdés, Fishman, Chávez, and Pérez 2008; Yeh 2004). Immigrant parents as educators usually adopt direct speaking methods and create opportunities for the children to use the heritage language frequently. However, this relies on family resources and on whether the parent educators adopt appropriate methods of instruction. Inappropriate instruction such as rigorous instruction may actually have a greater impact on the abandonment of home languages by children (Valdés *et al.*, 2008).

Heritage language learning seems to be more challenging for the cross-national marriage family, in which one of the parents is a first-language speaker and the other is an immigrant who is responsible for transmitting the heritage language. There are 498,368 married immigrants in Taiwan (about 2.1% of the total population). Sixty-eight percent of them are from the People's Republic of China (PRC), and remaining from Southeast Asian countries (mainly Vietnam 56.4%, Indonesia 17.5%, and Thailand 5%) (National Immigration Agency, 2015). Heritage languages (the languages used in Southeast Asian countries such as Thai, Vietnamese, Indonesian language, *etc.*) are seldom used in cross-national marriage families. Only 1.6%-2.8% of the parents in cross-national marriage families speak Southeast Asian languages at home, as indicated by the results of Yeh's investigation (2015). Mandarin (first language) or Taiwanese (local dialect) dominate in most cross-national marriage families.

Moreover, in Taiwan, the average level of education for immigrant mothers from Southeast Asian countries is primary school or lower-secondary school education (National Immigration Agency, 2010). Additionally, some parents have part-time jobs in the evening. Finally, Language learning materials for cross-national marriage families are also limited (Yeh, 2004). These factors create difficulties for heritage language learning at home.

With the development of globalization, multiple language capability has become a critical future requirement for people. The Taiwanese government has considered ways to familiarize children from cross-national marriage families with their mother or father's first language in Southeastern Asia. Many programs such as 'The National Torch Project', which encourages children from cross-national marriage families to attend heritage language courses at school, have been conducted (National Immigration Agency, 2013).

Parents were regarded as the main heritage language transmitters to children in the past. However, this approach was hampered by limitations in parents' abilities and the lack of learning materials (Yeh, 2012). Firstly, only one parent, usually the mother, is a southern Asia heritage language speaker in most cross-national marriage families in Taiwan. In addition, the parent needs to learn Mandarin or Taiwanese. It is very difficult for her/him to teach the child her/his heritage language. Additionally, the parent might not be able to teach the child effectively enough to obtain an educational degree. Secondly, although there are some teaching programs available for schools to teach students the heritage language, there are few appropriate instructional materials.

In order to overcome the above-mentioned limitations and to enable children from cross-national marriage families to learn southern Asian heritage languages, one solution is providing children with self-learning materials. However, how are such self-learning materials to be developed? Adopting information communication technology (ICT), especially portable digital devices might be one approach to this issue. Integrating ICT in schools is an effective way to motivate children's academic achievement and reading (Yang, Yu, & Sun, 2013; Yang & Wu, 2012). Rapid educational and technological advances, including the digital environment created by mobile applications, are accelerating the demands of online learning objects (Meurant, 2011).

Appropriate digital learning devices and resources may assist students' new language learning (Meurant, 2010). A computer-assisted language learning (CALL) system can provide an electronic language learning environment that can enable learners with mother tongue deficiencies to enhance their knowledge of grammar and acquire writing skills (Katushemererwe & Nerbonne, 2015). Related studies have focused on English as a Foreign Language (EFL). For instance, Golonka, Bowles, Frank, Richardson, and Freynik (2014) show that technology can make a measurable impact on foreign language learning, particularly computer-assisted pronunciation training and automatic feedback. The successful integration of integrating technology and language learning in this study motivated the current study's use of apps with a feedback system to help children from cross-national marriage families in Taiwan to learn the heritage language of their immigrant parents.

Apps are becoming increasingly popular on tablet computers. An effectively designed self-learning app might inspire children to learn heritage languages, particularly minority languages that lack teaching materials and language teachers. Self-learning apps might overcome the problem of the shortage of language teachers and make the heritage language available when children are provided with portable devices. If bilingual picture book apps are successful, this would not only benefit the children in Taiwan but also children in similar situations in other countries such as the USA, Germany, and the UK, which have high percentages of immigrants (OECD, 2013).

However, no previous study has examined the effectiveness of apps in assisting children of immigrant families in learning heritage languages. The current study aimed to address this gap by developing digital picture book apps to explore their effects on assisting children to learn heritage languages independently. We chose to adopt picture book apps because picture books are an effective tool in bilingual education (Godwin-Jones, 2011; Huang, Baek, & Cho, 2012; Meurant, 2011). Picture books are also the most powerful and personal way to learn about the world (Huang *et al.*, 2012). Therefore, as picture books enable both cross-cultural understanding and language learning, they can be used in multicultural contexts.

As picture books are expected to promote self-learning, they should use the first language that children are familiar with as the prompt. Accordingly, this study developed picture books using a bilingual approach. In doing so, the impact of children's proficiency in the first language on heritage language learning was also explored.

Given the importance of heritage language learning and the potential solutions for facilitating this learning, the aims of this study were two-fold. First, bilingual tablet picture books apps were designed to encourage children's self-learning. This study also evaluated the effectiveness of the tablet apps on heritage language learning. Concrete research questions are provided below:

1. Is a tablet bilingual picture book app an effective tool to facilitate pupils' heritage language learning?
2. Do children's abilities in the first language benefit their heritage language learning?

DESIGNING PICTURE BOOK APPS

Godwin-Jones (2011) suggested that tablets would be a popular product in the near future, and having language learning applications would be highly desirable if the target audience had a marked predominance in one platform. We designed bilingual tablet picture books apps to assist in bilingual learning and promote Vietnamese language learning for children from cross-national marriage families because immigrant mothers from Vietnam are in a majority in Taiwan.

The tablet picture books were designed based on the principles of 'universal design for learning' (UDL) (CAST, 2011). UDL encourages learners to explore the materials according to their own learning interests. The UDL is a set of principles, multiple representations, engagements, and expressions for curriculum development that give all individuals equal opportunities to learn (CAST, 2011). It is not a single, one-size-fits-all solution, but rather a set of flexible approaches that can be customized and adjusted for individual needs.

Tablet picture books use multiple forms of visual and audio presentations for the vocabulary items in the book. The users are no longer passive receivers because they can select multiple tablet functions to interact with the learning materials. The tablet picture book apps were developed on the Android operating system.

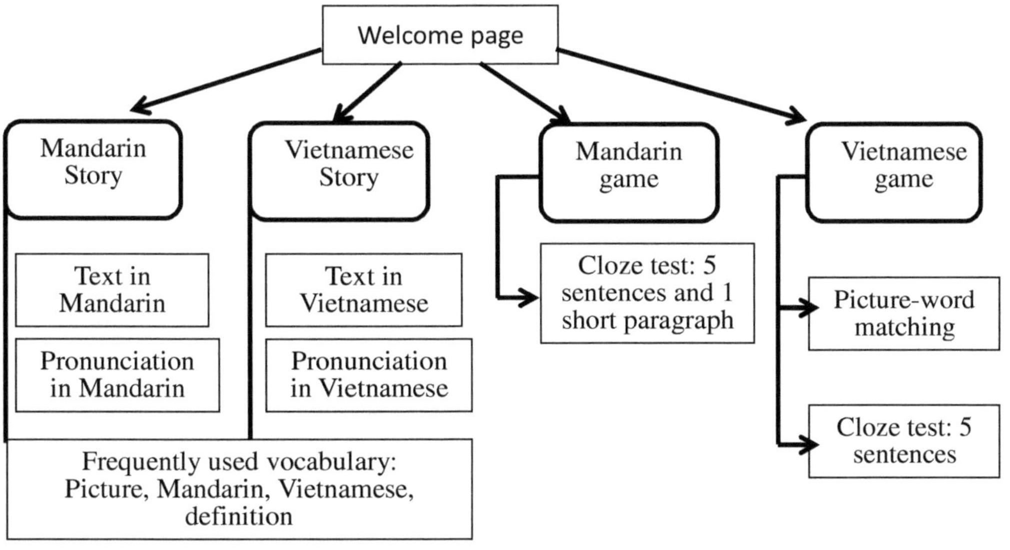

Fig. (1). Structure of the tablet storybook, *Changeful Mothers.*

Acquiring high frequency vocabulary might be more effective for learning a second language. For children whose parents – either the mother or father – are from Vietnam, Vietnamese is like a new second language because of the lack of opportunities to learn the language. We assumed that such children would be encountering Vietnamese for the first time. Therefore, the picture books included a story written with Mandarin vocabulary items frequently used in the first and second grades in Taiwan (Chen, Ko, & Chen, 2011). Key vocabularies were marked in red. Their definitions and pronunciations could be read when users touched the highlighted vocabulary items.

The framework of the picture book app was developed using a fixed structure. The framework is shown in Fig. (**1**). Along with the welcome page, the tablet picture book contained four parts (Fig. **1**): a story in Mandarin (including text and reading), a story in Vietnamese, Mandarin games, and Vietnamese games.

In addition to reading paragraphs aloud in Mandarin and Vietnamese, the key vocabulary items could be pronounced in Mandarin and Vietnamese, and be explained through multiple presentations (picture, voice, and text). The app also provides a picture-vocabulary matching game that allows children to match the picture with the Vietnamese vocabulary. This helps the children learn the meaning of the Vietnamese vocabulary.

Two picture book apps using the framework mentioned above were developed. The names of these books are *Changeful Mothers* (Fig. **2**) and *Phở* (a traditional Vietnamese noodle) and *Vermicelli* (a traditional Taiwanese noodle). Screenshots of pages, including multiple representations of the target vocabulary, and a vocabulary-picture matching game, are shown in Fig. (**3**). For the purpose of the experiment in this study, two print-based picture books were created; the same content was included in the print books in colors but without the games.

METHODOLOGY

In order to evaluate the effect of the tablet picture books on participants' heritage language learning, an equal group experiment was adopted (Fig. **4**). Eight undergraduate students that were majoring in early childhood education were trained to be research assistants to collect experimental data in this study.

Fig. (2). Screenshot of a bilingual tablet storybook, *Changeful Mothers.*

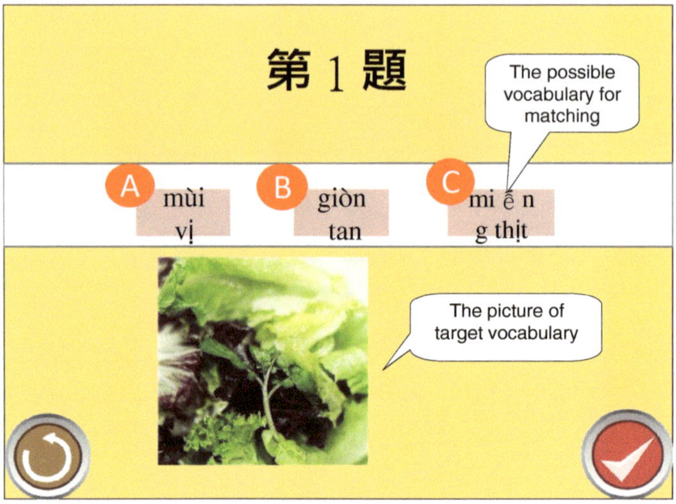

Fig. (3). The Vietnamese vocabulary-picture matching game in the tablet.

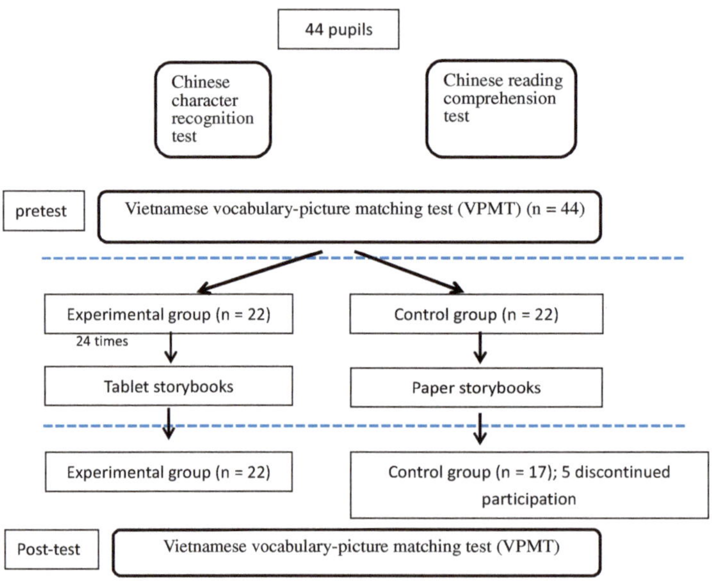

Fig. (4). Experimental process.

Participants

Forty-four 5th or 6th grade pupils from cross-national marriage families with Taiwanese fathers and Vietnamese mothers were recruited to participate in this study after gaining their parents' consent. None of the participants had learned Vietnamese before this study. They were assigned to the experimental or control groups randomly. However, five pupils in the control group withdrew because they felt that reading printed picture books was 'boring'. Finally 39 participants, 18 boys and 21 girls (mean age = 11.3 years) finished the experiment; 22 were in the experimental group and 17 in the control group. The fifth- and sixth-grade pupils were selected because they are more familiar with tablets. They have had at least two years of experience learning English in primary schools (one hour per week). Hence, reading Vietnamese vocabulary words would be easier for them. All the participants in experiment group were confirmed to be familiar with operating tablets before entering the experiment.

Experiment Design

An equal group experiment was conducted with a two-factor ANCOVA when literacy capabilities were unequal between the two groups. Although the participants were assigned into the experiment or control group randomly, it was difficult to equalize the language capabilities between two groups. Accordingly, the capabilities of Chinese character recognition and Chinese reading comprehension were adopted as co-variances to control their impact on Vietnamese learning if they were unequal.

The independent variable of this study was the condition of reading a printed picture book or a picture book app. The dependent variable of this study was the performance of Vietnamese vocabulary learning, which was measured by matching pictures with Vietnamese vocabulary.

In order to control for consistency during the experiment, the experiment was conducted in a quiet school classroom instead of in participants' homes. This also allowed us to ensure that the participants read through the picture books by themselves.

Instruments

The participants were required to take the Chinese Character Recognition Test (CCRT) (Huang, 2008) and the Reading Comprehension Screening Test (RCST)

(Ko, 1999) before the experiment. CRTT contains 200 frequently used Chinese Characters (or Hanja). The internal consistency of the test is .99, coefficient of stability ranges from .81-.95, and the split-half reliability is .99. In terms of validity with Chinese performance scores, the correlation coefficient ranges from .36–.76; the validity with the Chinese standard test is .48-.67. This is a good tool to identify participants' Chinese literacy levels.

RCST is a tool used to detect reading comprehension difficulties of primary school pupils from the second to sixth grades in Taiwan. The test includes 'word recognition' and 'comprehension' tasks. The internal correlation ranges from .36-.70; the internal consistency ranges from .75-.89; and the validity with other similar Chinese reading comprehension tests is significant.

The Vietnamese vocabulary-picture matching test (VPMT) designed by the researchers was used as the pre-test and post-test. The VPMT includes a vocabulary pool with 60 key vocabularies shown in the picture books. These test questions in Chinese were judged by four primary school teachers as suitable for first- and second-grade pupils to ensure that the vocabulary was not too difficult for the new learners of Vietnamese. The test questions were then translated into Vietnamese. Each question contains one picture and three Vietnamese vocabulary options.

The pre- and post-VPMT included 25 questions that were randomly selected from the vocabulary pool. The pre-test and post-test were in the form of a question sheet. The participants circled the correct answer with pencils. They had 30 minutes to finish the paper test. The research assistant graded the tests and recorded the results of each participant.

The Asus TF 700 Transformer pad was used to conduct the experiment. The device is a 10.1-inch pad with an Android 4.0 operating system that is commonly used in Taiwan.

Procedure

All the recruited participants took three tests to confirm their Chinese reading and comprehension abilities in advance. As mentioned above, the tests included the

Chinese Character Recognition Test (CCRT) (Huang, 2008), the Reading Comprehension Screening Test (RSCT) (Ko, 1999), and the pre-VPMT.

The participants in the experiment group were taught to operate the tablet individually in the first session to make sure they could operate the picture book app correctly. (All the participants in the experiment group were familiar with operating the picture book app in the first session.) Then, they were asked to read the picture book app in the following steps. First, they were asked to read the Vietnamese picture book. Then, they were asked to play the Vietnamese games. Finally, they were asked to take the post-VPMT. There was no time limitation for reading the picture book apps.

The pupils in the control group read the printed storybook in Vietnamese without assistance from university students. After they finished reading, the participants took the VMPT paper test.

Data Analysis

SPSS 17.0 was used to analyze the data. An ANCOVA was conducted when literacy capabilities were unequal between the two groups. Otherwise, a *t*-test was used to test the effect of the intervention.

RESULTS

As Table **1** shows, since Chinese literacy capabilities were unequal between the experimental and control groups (CRTT: $t = 2.92$, $p = .00$; RCST: $t = 2.16$, $p = .04$; VPMT: $t = -.11$, $p = .91$), a two-factor ANCOVA was conducted to examine the effect of tablet picture books on heritage language learning.

Table 1. The mean and standard deviation of Chinese literacy and Vietnamese vocabulary ability for the experiment and control group.

CRTT				RCST				VPMT			
EG[a]		CG		EG		CG		EG		CG	
Mean	SD	Mean	SD	Mean	SD	Mean	SD	Mean	SD	Mean	SD
54.64	6.85	46.65	10.23	75.23	16.86	61.76	21.64	7.09	4.15	7.24	3.71

Note. a: EG represents experiment group, CG represents control group.

The results of the ANCOVA test are shown in Table **2**. The results indicate that the experimental group (EG) had higher VPMT scores than the control group (CG) ($F(1,35) = 4.77, p = .04$; Cohen's d = 1.17) after excluding two co-variances (CRTT, RCST).

Table 2. The results of ANCOVA for the intervention of tablet picture books.

Source	SS	*df*	Mean Square	*F*
CCRT pre-test	20.17	1	20.17	1.68
RCST pre-test	23.89	1	23.89	1.99
Group	57.34	1	57.34	4.77
Error	420.36	35	12.01	

The second research question proposed in this study was whether tablet picture books enhanced Vietnamese learning for all participants or only those who had higher levels of Chinese proficiency. The 22 participants in the experimental group were divided into two groups: one with higher CCRT scores ($> 52.00, n = 12$) and the other with lower CCRT scores ($< 52.00, n = 10$). The participants with higher levels of Chinese also showed higher VPMT scores ($M = 14.5$, SD=4.25), compared with the low CCRT group ($M = 10.7$, SD=3.89) ($t = 2.19, p = .04$, Cohen's d = 0.93).

DISCUSSION AND CONCLUSION

This study found that tablet picture book apps benefit children's heritage language learning. The results demonstrated that picture book apps have an effect on Vietnamese learning for Taiwanese pupils from cross-national marriage families. In particular, those pupils who had better capabilities in Chinese character recognition were able to learn better. Korat (2010) concluded that ebooks provide support for young pupils' vocabulary acquisition and word reading. This study showed that the application of ebooks to heritage language learning is also effective.

Heritage language learning process used to be perceived as a sociocultural process that took place at home (He, 2010). A number of immigrant parents preferred that their children integrate themselves into their host society (Höhne and Koopmans,

2010; Moin, Schwartz, and Breitkopf, 2011), although gaining multiple language capability was considered useful for the youth in globalized society, heritage language learning or cultural transmission was not a primary concern for immigrant parents.

This study demonstrated that picture book apps can encourage self-learning of heritage languages. Moreover, picture book apps can be downloaded and shared easily through tablets or smart phones. This study illustrated alternative ways that children can effectively learn their heritage language by themselves.

In addition, we found that learners with better first language capabilities could acquire heritage language vocabulary more effectively. Song and Fox (2008) used user-determined devices in the study and concluded that EFL students' vocabulary building activities became powerful when coupled with the students' communication with peers. Dialogue and sociocultural interaction between children and immigrant parents could also be some factors responsible for this. It appears that children with better first language ability do not find it difficult to transmit their experiences of first language to the heritage language through self-learning.

This study sheds light on the application of tablet picture books to heritage language learning by children from cross-national marriage families. The results can also be applied to bilingual learning. Tablet picture book apps can be a supplement for bilingual or multilingual learning and encourage a learner-centred and self-directed mode. Tablet picture book apps represent a supportive and easy-to-operate device that can help children increase their heritage language vocabulary and reading.

One important limitation about the picture apps adopted in the current study is that they provide only vocabulary learning. Future apps may include syntax learning, listening comprehension, and speaking to meet the full requirements of language learning (Godwin-Jones, 2011). Further, pupils who have a lower first language performance may need more assistance in heritage language learning. Identifying the appealing parts of picture book apps may be useful here. As this study has shown, picture book apps, in addition to immigrant parents' sociocultural interactions, act as a successful self-learning supplement. The next step is to encourage children to learn heritage languages by themselves through the use of language apps.

CONFLICT OF INTEREST

The authors confirm that the authors have no conflict of interest to declare for this publication.

ACKNOWLEDGEMENTS

This work was partially supported by the Ministry of Science and Technology in Taiwan (NSC 101-2420-H-415-001). The authors thank the research assistants and primary school pupils who collaborated with them for data collection.

REFERENCES

CAST (2011). *Universal design for learning guidelines Version 2.0.* Wakefield, MA: CAST.

Chen, M.C., Ko, C.C., Chen, S.W. (2011). *Chinese text entry training system developed based on core vocabulary approach.* Asia-Pacific Conference on Technology Enhanced Learning, Xi-an China

Godwin-Jones, R. (2011). Emerging technologies mobile apps for language learning. *Language Learning and Technology, 15*(2), 2-11.

Golonka, E.M., Bowles, A.R., Frank, V.M., Richardson, D.L., Freynik, S. (2014). Technologies for foreign language learning: A review of technology types and their effectiveness. *Computer Assisted Language Learning, 27*(1), 70-105.
[http://dx.doi.org/10.1080/09588221.2012.700315]

He, A.W. (2010). The heart of heritage: Sociocultural dimensions of heritage language learning. *Ann. Rev. Appl. Linguist., 30*, 66-82.
[http://dx.doi.org/10.1017/S0267190510000073]

Höhne, J., Koopmans, R. (2010). Host-country cultural capital and labour market trajectories of migrants in Germany. *The Impact of Host-Country Orientation and Migrant-Specific Human and Social Capital on Labour Market Transitions.* WZB Discussion Paper No 2683.

Huang, J., Baek, Y., Cho, J. (2012). Technology-supported storytelling (TSST) strategy in virtual world for multicultural education. *Communications in Computer and Information Science, 341*, 101-108.
[http://dx.doi.org/10.1007/978-3-642-35248-5_15]

Huang, K.L., Chen, K.S., Ho, C.H. (2013). Promoting in-depth reading experience and acceptance: Design and assessment of tablet reading interfaces. *Behaviour and Information Technology, 33*(6), 606-618.

[http://dx.doi.org/10.1080/0144929X.2012.759625]

Huang, S.H. (2008). *Chinese character recognition test.* Taipei, Taiwan: Psychological Publisher.

Katushemererwe, F., Nerbonne, J. (2015). Computer-assisted language learning (CALL) in support of (re) - learning native languages: The case of Runyakitara. *Computer Assisted Language Learning, 28*(2), 112-129. [http://dx.doi.org/10.1080/09588221.2013.792842]

Ko, H.W. (1999). *Reading comprehension screening test.* Taipei, Taiwan: National Science Council. (in Chinese)

Korat, O. (2010). Reading electronic books as a support for vocabulary, story comprehension, and word reading in kindergarten and first grade. *Computers and Education, 55*, 24-31. [http://dx.doi.org/10.1016/j.compedu.2009.11.014]

Meurant, R.C. (2010). The iPad and EFL digital literacy. *Signal Processing and Multimedia Communications in Computer and Information Service, 123*, 224-234.

Meurant, R.C. (2011). Developing EFL/ESL cognitive structures using a mobile app to exploit a spatial morphology of verb forms in simple sentences. *International Journal of u- and e- Service Science and Technology, 4*(4), 63-74.

Moin, V., Schwartz, M., Breitkopf, A. (2011). Balancing between heritage and host languages in bilingual kindergarten: Viewpoints of Russian-speaking immigrant parents in Germany and in Israel. *European Early Childhood Education Research Journal, 19*(4), 515-533. [http://dx.doi.org/10.1080/1350293X.2011.623530]

Montrul, S. (2002). Incomplete acquisition and attrition of Spanish tense/aspect distinctions in adult bilinguals. *Bilingualism: Language and Cognition, 5*(1), 39-68. [http://dx.doi.org/10.1017/S1366728902000135]

National Immigration Agency (2010). *The survey of foreign spouses' living condition and needs from Southeast Asian countries and Mainland China in 2008. Retrieved from: http://www. immigration.gov.tw/lp.asp?ctNode =31540&CtUnit=17111&BaseDSD=7&mp=1.*

National Immigration Agency (2013). *The National Torch Project for the cross-national marriage family. Retrieved from: http://www.immigration.gov.tw/np.asp?ctNode=33976&mp=1.*

National Immigration Agency (2015). *Numbers of foreign spouses from Southeast Asia and PR China. Retrieved from: http://www.immigration.gov.tw/ct.asp?xItem=1247109&ctNode=29699&mp=1.*

Nomura, T., Caidi, N. (2013). Heritage language acquisition and maintenance: home literacy practices of Japanese-speaking families in Canada. *Information Research: An International Electronic Journal. Retrieved from: http://InformationR.net/ir/18-3/colis/paperC37.html, 18*(3)

OECD (2013). *Inflows of foreign population into selected OECD countries and the Russian Federation. Retrieved from: http://www.oecd.org/statistics/.*

Park, H., Tsai, K.M., Liu, L.L., Lau, A.S. (2012). Transactional associations between supportive family climate and young children's heritage language proficiency in immigrant families. *International Journal of Behavioral Development, 36*(3), 226-236. [http://dx.doi.org/10.1177/0165025412439842]

Park, S.M., Sarkar, M. (2008). Parents' attitudes toward heritage language maintenance for their children and their efforts to help their children maintain the heritage language: A case study of Korean-Canadian immigrants. *Language Culture and Curriculum, 20*(3), 223-235.

Song, Y., Fox, R. (2008). Using PDA for undergraduate student incidental vocabulary testing. *European Association for Computer-Assisted Language Learning, 20*(3), 290-314.

Spolsky, B. (2011). Does the United States need a language policy? *CAL Digest. Retrieved from: http://www.cal.org/resources/digest/digest_pdfs/does-the-us-needs-a-language-policy.pdf.*

Tseng, V., Fuligni, A.J. (2000). Parent adolescent language use and relationships among immigrant families with East Asian, Filipino, and Latin American backgrounds. *Journal of Marriage and Family, 62*, 465-476. [http://dx.doi.org/10.1111/j.1741-3737.2000.00465.x]

Uttal, L., Han, C.Y. (2011). Taiwanese immigrant mothers childcare preferences: socialization for bicultural competency. *Cultural Diversity and Ethnic Minority Psychology, 17*(4), 437-443. [http://dx.doi.org/10.1037/a0025435] [PMID: 21988582]

Valdés, G., Fishman, J.A., Chávez, R., Pérez, W. (2008). Maintaining Spanish in the United States: Steps toward the effective practice of heritage language re-acquisition/ development. *Hispania, 91*(1), 4-24.

Yang, X., Yu, S., Sun, Z. (2013). The effect of collaborative annotation on Chinese reading level in primary schools in China. *British Journal of Educational Technology, 44*(1), 95-111. [http://dx.doi.org/10.1111/j.1467-8535.2011.01277.x]

Yang, Y.T., Wu, W.C. (2012). Digital storytelling for enhancing student academic achievement, critical thinking, and learning motivation: A year-long experimental study. *Computers and Education, 59*, 339-352. [http://dx.doi.org/10.1016/j.compedu.2011.12.012]

Yeh, Y.C. (2004). *The study of the welfare and living condition of the international marriage family and their children in Tainan City.* Tainan, Taiwan: Tainan City Government.

Yeh, Y.C. (2012). A review of the execution of the mother-language policy among new immigrants. *Educational Resources and Research, 106*, 57-81.

Yeh, Y.C. (2015). The Southeast Asian heritage language policies in Taiwan. In: Wen, S.M., (Ed.), *New visions for national education: Reference, change and innovation* (pp. 53-72). Taipei, Taiwan: National Academy for Educational Research.

CHAPTER 11

Issues of Communication Behavior on Using Internet

Su Chong Chuang[*]

Graduate Institute of Public Policy, National Chiayi University, Chiayi, Taiwan, R.O.C.

Abstract: Due to the Internet, a virtual platform filled up with networks, boundaries between people have become a lot less, the way people communicate and interrelate is changed. However, conflicts are embedded in interactions through electronic networks, while the new social norms are emerging and yet not legally nor ideally launched. Based on observations of hot issues in Taiwan, the author sees that the cyber world is full of moral dilemmas and value conflicts. Demanding for anonymity and online free expression may significantly characterize the virtual world from the real world, the moral development of Internet users in cyberspace shows a need for further studies.

Keywords: Communication behavior, Cyber addiction, Cyber bullying, Cyber pornography, Ethics, Internet, Interpersonal relationship, Moral development, Online expression, Online manhunt, Online user review.

INTRODUCTION

The last century has paved an astonishing way for the most revolutionary computer techniques to come, among which the information and communication technologies (ICTs), specifically the Internet, have a critical impact on people's life. According to the survey by the Internet World Stats in 2015, it is about 84% of Taiwan's population who are Internet users, while the Facebook subscribers on December 31, 2012 were about 56.5%. People are increasingly relying more on the Internet for daily lives. They work, consume, sell, do the banking, take

[*] **Corresponding author Su Chong Chuang:** Graduate Institute of Public Policy, National Chiayi University, Chiayi, Taiwan, R.O.C; Tel/Fax: 011886-5-2717977; E-mail: make2win@gmail.com

Kuo Hung Huang (Ed.)

classes, do the leisure, and communicate on line. By the Internet, a virtual platform filled up with networks, societies are tremendously influenced, lives are overwhelmingly impacted, boundaries between people have become less, the way people communicate and interrelate is changed. However, conflicts are embedded in interactions through electronic networks, while the new social norms are emerging and yet not legally nor ideally launched.

The electronic information systems, specifically the Internet either through smartphones or computers, are bringing us more sources for information and are changing the social networking. Nikou and Bouwman (2014) mentioned that people have been using computer as a communication tool to build online virtual community, and that people are using social network services *via* mobile devices, specifically the smartphones. Data from the most popular social network, Facebook, shows that its daily active users (DAUs) were 968 million on average for June 2015, representing an increase of 17% year-over-year. Mobile DAUs were 844 million on average for June 2015, representing an increase of 29% year-over-year (Facebook Report, 2015). The rate of social network users is dramatically increasing. Tencent QQ, the largest and most used internet service portal in China, revealed its smart device monthly active user (MAU) of QQ was 603 million, representing an increasing of 23% year-over-year, and its combined MAU of Weixin and WeChat were 549 million, representing an increase of 39% year-over-year (Tencent, 2015). Renren (2015), also in China, a leading real-name social networking internet platform, had approximately 223 million activated users by December 31, 2014. These mobile social network services are absolutely changing the way people communicate and interrelate.

The Internet has sped everything up, and it has made the world less boundary in all dimensions. However, the virtual online environment is much harder to control. Advantages of getting on the World Wide Web may be a lot, the risks of getting online could never be less. Socially, people may get their social connection instantly *via* Internet, but they also could get mistreated or victimized due to some Internet users' improper behavior. More and more businesses and industries have heavy usage of computer mediated communications, these electronic commercial activities have been creating paradigm shifts in business management and in business performances (Lagraña, 2010). The Internet contributes to a momentous shift in world economy, and the political and social arena as well. A great number of researches (Tolbert & Mcneal, 2003; Kenski & Stroud, 2006) put their concern on the political consequences of using the Internet, particularly in aspects of political participation and civic engagement. Impacts of Internet are overwhelmingly spread out, this article is to focus on the communication behavior by using Internet, specifically ethical issues raised in between.

It has been more than 20 years since Margaret Lynch (1994) had her research on ethical issues in electronic information systems. Lynch noted the importance of ethics because she saw the power in information and the legal lag way behind new techs. Lynch believed the networks have exciting possibilities and they were sources of power that might threaten some systems. She also found networks to be social places enabling more social interactions. According to Lynch, the virtual anonymity of Internet allows more communication with users' characteristics hidden on one hand; on the other, it allows interaction without shared responsibility.

Lynch (1994) pointed out the issues of using electronic information systems include (1) standards of civility for rights and responsibilities in distributing information, (2) the access to information, (3) privacy, (4) misuse of data, and (5) international considerations. Lynch extended these issues to electronic networks, electronic databases, and geographic information systems. For each arena, Lynch believed ethical decisions are required for problems.

INTERNET USE IN TAIWAN

By the end of 2013, the population of Internet users was close to 18.6 million, which stood for an 80% penetration rate out of the 23.35 million total population. According to a 2014 report from Taiwan Network Information Center (TNIC), Taiwan's Internet growth rate reached 77.6% in 2014, and that the number of Wi-Fi and mobile Internet users above 12 years old reached 12.6 million. The TNIC also had a report in 2014 telling that social networking was the most popular smartphone activity (64%), with a highest social media penetration rate (69%) in the region.

Most Debated Social Events

Taiwan is highly ranked in terms of press freedom, social debates wouldn't have a chance to escape from the press coverage. The author collects the most debated Internet related cases in the past 5 years from UDN News Database (2010-2015), which is so far believed to be the most complete data set with big data. Cases are cited based on key words: Internet, Internet dispute, Internet event. The top 10 most-often-discussed Internet related cases contain events and phenomenon as the followings:

(a) Online bullying led a popular young entertainer to commit suicide resulting in a massive public debate; (b) Smart phones were raising the numbers of cyber-bullying cases; (c) A big data showed nearly 40% of the youth who play online games have the experiences of being cyber-bullied; (d) Samsung was fined for paying hired writers and designated employees to post fake web reviews in online forums praising Samsung and criticizing competitors; (e) Netizens or Bloggers' sharing experiences of products, especially the cosmetics, were identified as advertising or exaggerated promotion; (f) With a local celebrity Lee posting photos on her Facebook page showing off her entering the cockpit of an Apache helicopter, five military officials involved were later penalized; (g) The Formosa Fun Coast related- An online message declared that the Democratic Progressive Party has been quiet about the water park tragedy because of its financial links to the party; (h) A man posted on his Facebook about MERS contagious diseases, but later the FDA discredited the rumor. The man regretted posting unfounded statements; (i) With its digital mobilization strategies by using Internet social media, the Sunflower Movement represents a new era in Taiwan's social movements; (j) With the help of manhunt, lots of criminal cases were traced down. Taiwan has made the "Human Flesh Search" legal in cases of public interest.

Communication Behavior Issues Raised

1. A celebrity's suicide caused an overwhelming blame on cyber-bullying. There is an urgent demand from the public asking for self-regulation of social networking sites, and a heated debate over how the law should have protected the citizens from being cyber-bullied.

2. The occurrence of cyber-bullying has increased with the prevalence of smartphones. With smartphones, people become even more easily targeted by online predators, especially the young generation. They could be either bullied or sexually harassed by people they know or not know in real life.
3. A large percent of people trust online user reviews and lots of people are affected in their consuming decisions by these reviews. Businesses, such as hotels, restaurants, cosmetics, have been relying heavily on web-based reviews resulting a mass astroturfing, Samsung was just one of the companies. Most Consumers are not capable of judging whether comments posted online are real or fake. Online reviews paid for marketing becomes an issue of Internet ethics.
4. Rumors could be everywhere in the virtual world for certain purposes and be broadcasted even faster than they are in a real world. Lots of messages are reprinted on online social medias, such as "line" or "facebook", alleging some information leading all the time to intervention and investigation from the police. Ethical concerns arise from those rumors on social media. Internet communities are capable of amplifying effects of ethical or unethical judgments.
5. For people posting personal messages on Facebook or on other social media sites that cause disturbances, aren't they grabbing too much public attention and discussion on personal individual behavior? Could it be a waste of social sources?
6. Behaviors of cyber manhunts or digging up personal information trigger issues about online invasions of privacy and expose the problem of protection of personal information.
7. The use of online materials makes the young generation and the educational community on campus vulnerable to infringe copyright. If forwarding emails may possibly constitute a copyright infringement, then what about the transmission of messages and data online *via* social media? Will the net users become the highly risky and targeted groups to break the law?

Other than those ethical issues above, what has been bothering and worrying the Internet users and nonusers also include issues of cyber-safety, cyber-addiction and cyber pornography, to name a critical few. The use of Internet becomes a "can't live without" activity for a lot of people in Taiwan. According to a survey by the Internet World Stats in 2015, it is about 84% Taiwan's population that are Internet users, while the Facebook subscribers on December 31, 2012 is about 56.5%. The use of Internet is not only indispensable but excessive to some of the users, and become problematic therefore. Lots of researches identify the excessive Internet use as a pathological or addiction of Internet use. By 2011, based on previous researches, Ma (2011) grouped Internet Addiction into three major concepts. Among them, Ma suggested the addiction could be technological, as was proposed by Griffiths (2000) that the Internet might just be a "means to fuel their other addiction and interest".

Scenario 1: A Closer Interpersonal Relationship?

Back in 1991, Faith Popcorn in her best-selling book, predicted a "cocooning" age to come, foreseeing that people would be too afraid to go out. She might not be right on the "too afraid" saying, but she was, in some "physical" perspective, right that people are "cocooning" in their residences. In the late 20th century, given the emergence of electronic network with the development of high techs, humans are entering a new era. Cocooning or the SOHO life style were just a starting, the digital individual with virtual self has become the main role fulfilling the Cyberspace, virtual community are taking critical and more places of the face-to-face interpersonal communication.

Researches about the impacts of Internet use are a lot, but their findings vary. Variables investigated include information facilitation, emotional support, psychological well-being, communication, extraversion, perceived social support, social involvement, computer skills, and closeness of friends *etc.* (Fung *et al.*, 2001; Chang, Choi, Bazarova, & Löckenhoff, 2015). For Caplan (2003), his finding suggested negative effects of the internet use on well-being. However, Campbell *et al.* (2006) found positive effects for similar variables.

For younger generation, Lin, Zhang, Jung, and Kim (2013) showed a data in the United States that 93% of adolescents aged 12-17 were online in 2009; while in East Asia, the Internet penetration of adolescents in Korea, Taiwan, and Singapore had approached almost 100% by 2007. Rideout, Foehr and Roberts (2010) noted that American children 11 to 14 years old spend most of their online time on social networks. Livingstone, Haddon, Gorzig, and Ólafsson (2011) found 38% of UK children 9 to 12 years old have a profile on one or more social networks. In May, 2011, Consumer Reports noted, "of the 20 million minors who actively used Facebook in the past year, 7.5 million of them were under 13". The cognitive and emotional development of the younger generation may have being shaped by online social networking activities. Blackwell, Lauricella, Conway and Wartella (2014) found that "Video clip and movie Web sites, especially YouTube, and social networking sites, in particular Facebook, are among the favored Web sites of 8- 12-year-old Internet users, particularly for older children". Blackwell *et al.*, (2014) contended there is a lack of age-appropriate content for 10- to 12-year-olds to meet their social, emotional, and cognitive developmental needs. By finding that "children are still finding their way to Web sites that are not considered age-appropriate", Blackwell *et al.*, (2014) showed their concern on the high incidence of underage children on social networking sites, they found that many underage children go on Facebook by lying about their age.

Boyd, Hargittai, Schultz, and Palfrey (2011) found that many parents are aware and allow their children to create accounts on Facebook with fake ages and profiles. Blackwell *et al.* (2014) believed that these behaviors could have important developmental implications. They found that "engaging in thee online social interactions prior to necessary cognitive and emotional development that occurs throughout middle childhood could lead to negative encounters or poor decision-making". They also pointed out that "younger children favor social networking sites, which are traditionally reserved for teens and young adults, their interactions on these sites may speed up their social development at an emotional and cognitive stage that may not be ready for advanced social development tasks". Blackwell *et al.* urged the "need to provide media literacy at an earlier age to ensure that children engage in safe and developmentally appropriate interactions and activities online as well as a need to develop more engaging online material for youth in middle childhood". Children and adolescents are born in the Internet age, they spend huge amount of time to interact with the world on line. Researches support the view that the younger generation is more susceptible to online virtual world.

According to Chang *et al.* (2015), population of online social network sites is growing, the reason could be that the networks may either fulfill people's social needs or making more interpersonal ties, or facilitate information sharing (Southwell, 2013; Chang *et al.*, 2015). Even though those networks enable tremendously large relation nets, scholars like Ellison, Steinfield and Lampe (2007) believed some of the online relationships are weak, superficial, or even ephemeral. According to Ellison *et al.* (2007), for a Facebook user, about three-fourths of the user's friends are friends without "common personal history and meaningful connections". In other words, only one-fourth of the user's friends are actual friends. However, for college students, the proportion of such superficial relations is growing up (Manago, Taylor, & Greenfield, 2012). Looking back to what Young (1998) had observed from his classification of Internet addiction behaviors, he noted the addicts involved heavily in online relationships were having a "cyber-relationship addiction". Generally and obviously, with time goes by, people's social behavior is positively associated with their increasing use of Internet. Virtual communities bring users wide relation nets for social communication, representing more online connections with somehow not-so-strong and not-so-healthy but vague interpersonal relationships.

Scenario 2: The Cyber Addiction

Nowadays, more applications software are developed, more social web sites are built, and they are easily accessible, which results in more time spent by people in using Internet in daily life. Although positive outcomes such as knowledge sharing, instant messaging, larger net relations *etc.* may be associated with cyber communication, some users do experience unpleasant consequences.

According to Young (1998)'s grouping, the cyber-addicts also include: cyber-sexual addiction, net compulsions, information overload, and computer game addiction. Among those addictive behavior types, online computer games have successfully made themselves irresistible to a great percentage of Internet users, specifically the younger generation. Shek, Tang, and Lo (2008) had a survey of Chinese primary and secondary students and found one-fifth of them could be Internet addicted. Based on previous studies, Ma (2011) listed characteristics of Internet addicts as: excessive use of Internet, obsessive thoughts about the internet, pleasant feeling in Internet use, the tolerance symptom, diminished impulse control, the withdrawal symptom, negative impact on daily life, less parental and family interactions, less real friendship and romantic relationships, health problems, negative academic performance, and loneliness.

According to a nationally phone survey of 802 American parents and their 802 teens ages 12-17, Madden, Lenhart, Duggan, Cortesi and Gasser (2013) found that 78% of teens have a cell phone and 95% use the internet. A great percentage of teens, as was surveyed by Amanda Lenhart (2015), text their friends *via* communication platforms and online venues either occasionally or daily to keep friends in touch. Online games are always the sources that make the players addictive, and they are sophisticated and violent. Lenhart also found that gameplay is an important activity to establish and keep their friendships with others. However, one academic data (Taipei Times, 2012) showed that in South Korea, more than 2.4 million youngsters, or 30 percent of young people, had been suffering from the side effects of Internet and video game addiction. Internet addiction is thought to be a public health issue in South Korea. Besides, relationships formed based on gameplays could be weak and vulnerable, some even caused negative unintended effects (Zilberstein, 2013; Lenhart, 2015).

In Taiwan, the Taiwanese Society of Suicidology and the Taiwan Suicide Prevention Center announced their survey results telling the correlation between Internet addiction and emotional distress and suicide. According to the center, it was about 7 percent of people surveyed experienced negative moods within a week and 2.4 percent had a thinking to commit suicide. Besides, for those who were identified as having Internet gaming disorder, 24 percent of them appeared to be in effected by negative emotions and 11.5 percent considering suicide. The physical end of computer gaming addiction could be death, not to mention that there might be "back pain, eye strain, and carpal tunnel syndrome (Young, 1998)". In the past few years in Taiwan, there have been lots of death cases occurring in Internet café, where gaming marathons caused exhaustion and life threatening.

Lots of researches focus more on the problematic Internet use of younger generation, and most of their findings support a thought that children and adolescents are the age groups more likely to be addictive to Internet use at the expense of their time supposed to do sports or academic works. The more time they spend on line, the higher possibility they expose and involve themselves in a large variety of online activities, and become prone to exhibit addictive behaviors and the resulting outcomes.

Scenario 3: Cyber Pornography

Ever since the pornographic materials have been distributed on the Internet, the excessive growth of cybersex has made the Internet a media for pornography. Cyber-porn advertisements are absolutely widespread and unguardedly accessible; people exchange their sexual experiences, making their related sexual deals *via* the electronic net. Based on some meta analyses, a pervasive research finding was provided to suggest that consumption of pornography could result in serious effects both in attitudes and behavior (Lo & Wei, 2005) and in all ranges of ages.

Censoring the use of websites or Internet is authorized by some governments, however, such regulations for using Internet has raised debates over online ethics. In Young's (1998) study, cyber addicts included the cyber-sexual addiction that people spent a lot of time in adult websites for cybersex and cyber-porn.

Findings from Lo and Wei (2005) showed that about 40% of their surveyed Taiwanese adolescents had some exposure to Internet pornography, and "the exposure to Internet pornography relates to greater acceptance of sexual permissiveness and the greater likelihood of engaging in sexually permissive behavior". Furthermore, they found that "the relationships between Internet porn exposure and attitudes toward premarital and extramarital attitudes as well as sexually permissive behaviors were sustained".

Easy accessibility to free pornography online raises concerns from parents. Parents in Taiwan are worrying about their kids for living in a wired world that makes the younger age bracket exposed to pornography. While parents want to curb the availability of cyber-porn and keep the pornography business on Internet away from the youth, the legislation problem of reducing the dissemination of online pornographic materials and online expression is the impingement to the right of free speech. On one hand, Discussions have been issues whether it is appropriate and necessary to regulate the cyberspace; on the other, personal rights for free speech and privacy are demanded, as well as the development of industries on the Internet. In 2015, after the entertainer Peng committed a suicide, the debate was also once again triggered over laws on cyber-bullying and freedom of speech.

Scenario 4: Online Freedom of Expression

In Taiwan's most debated social events, a great percentage of those cases raise a correlated issue- expression online and the freedom of speech. Bullying or being bullied online has taken away many people's lives or ruined them. The entertainer's suicide was just one of the cases. In responding to the public's request for anti-cyber bullying and the Apache helicopter case, the Vice Premier Zhang noted in 2015 that the government won't engage new laws to limit online expression. The chairman of the National Communication Committee (NCC) also took the same stand in earlier announcement that the NCC will not regulate or suppress the online expression to ensure the right of free speech.

Through the Internet, more industries are paying for users to comment on their products. Web-based reviews become a dominant marketing strategy for some industries. In the case of Samsung, the Taiwan's Fair Trade Commission fined Korean Samsung for "astroturfing". The action of Samsung's hiring people to make fake web reviews was considered pollutants in social networks. According to *the Local Consumer Review Survey* 2013 of around 2800 local consumers in the United States, Canada and the United Kingdom, about 85% had read online reviews of local businesses, about 79% trusted online reviews as much as personal recommendations, and about 67% read 6 reviews or less before making a purchasing decision. Online astroturfing behavior has been too much for users to tell whether it's purely experience sharing or it's a paid-for comment. It was also found in Nielsen's 2012 report that consumers had more trust in online while making purchasing decisions, and had become more reliant on other consumers' reviews from social networks or consumer generated media. From moral perspective, consumers could be easily misled by unethical web reviews.

Other cases like personal information leakage, online rumors, election wording, and even online manhunt are either related to the right to free speech issue or the personal privacy issue. Accessing the Internet means stepping into the electronic world, in which users may mask their identities and exchange information with unlimited number of other users. The unlimited exchange of information brings concerns over the possibilities of conveying unappropriate materials online. Issues over accountability and responsibility deriving from information transmission and free expression online get increasing concern.

ETHICAL PERSPECTIVE

Considering what actions are right or wrong, or whether the behavior helps or harms on using Internet, the core concept is ethics. When Lynch (1994) identified Internet acceptable behavior, she believed it should contain simple standards of civility, which is one of the main components of ethics. Nearly 20 years back in 1998, Lynch noted some ethical issues that need to be concerned on Internet using: control of and access to information, privacy and misuse of data, and international considerations. Now, those ethical concerns do not fade out. Communication behaviors on Internet being good or evil, justice or crime are issues that need to be delineated from moral perspectives. Lawrence Kohlberg's (1984) theory of moral development was always used to examine social behavior. As was applied by Ma (2011) to examine the moral reasoning of antisocial Internet behavior, the Kohlberg's theory is also employed here for drawing the ethical pictures of the issues discussed above. Kohlberg's theory has six identifiable developmental stages for ethical behavior: (1) punishment and obedience; (2) individualism and exchange; (3) interpersonal relationships; (4) maintaining social order; (5) social contract and individual rights; (6) universal principles.

For people in the first two stages, it is the rewards and punishment to justify whether the behavior is right or wrong. In stage 3 and 4, people try to do duty, avoid blame, obey authority, and look for approval. Views from others do actually matter. People try to live up to social expectations and consider the society as a whole. For people at stage 5, differences between moral and legal right are recognized, different values, opinions, and beliefs are taken into account. At the last stage. People follow internalized principles of justice, no matter if they conflict with laws.

Given the anonymous convenience in a less regulated cyber world, net users' misconduct may not be easily caught or even defined. The authority seems to be too hidden to obey. From their perception, net users may not think that their free expression are bullying and killing some ones on Internet. In such cyber-bullying cases, boundaries are vague, no tangible authority to obey, no punishment to avoid, which is quite different from what K's stage 1 expected. A great number of net users exchange information, some of them even try to get much but pay less, leading the cyberspace to be filled with opportunistic hedonism, echoing what Kohlberg's stage 2 elaborated. For other stages, people were supposed to live up to what was expected and maintain the social order and follow some universal principles. However, in cyber world, unethical behavior like illegal downloading has been quite an issue for long. The manhunt issue explores another moral dilemma. Human flesh search is an action taken by net users to band together for tracking down people's true identities. By doing so, the searchers, out of an internalized justice perception, could be trying to maintain orders both in cyber and in real worlds. However, those behaviors are associated with the concern of violating privacy and the resulting danger to the society.

CONCLUSIVE NOTION

The cyber world is full of moral dilemmas and value conflicts. It is the anonymity and online free expression that significantly characterizes the virtual world from the real world. Given a less regulated and less law enforcement cyber environment, not many of the net users have actually taken full responsibilities or get punished for not obeying the social order in virtual communities, they might thus feel even freer to enter or create the dark side in cyberspace, where a lot of people keep losing or ruining their lives. Net users may have wider connection and ties to others by the fast spread-up of online social communities, yet the effects for better interpersonal relationship are still in debate. Net users are defending their positions and actions to justify their online behaviors based on beliefs such as: the paid-for comments are for businesses whether they are fake or not; manhunt is for justice even if the net users have to dig up personal information and infringe others' privacy. Kohlberg in his six stages of moral reasoning model contended that it is very rare to regress in stages, and stages cannot be skipped. However, moral conflicts and the resulting choice of behavior in cyberspace make Kohlberg's stages blurring. The thought comes here is that the moral reasoning stages by Kohlberg, the basis of ethical behavior, may need to be reconsidered for cyberspace by dividing them into two tracks- for the cyber world and the real world. Maybe the stages need a rotation- to progress through the stages in different order- for an alternative portray of the moral pictures of communication behavior on Internet in the cyber world.

CONFLICT OF INTEREST

The author confirms that the author has no conflict of interest to declare for this publication.

ACKNOWLEDGEMENTS

This article is an individual effort. Still, I would like to show my great appreciation to Professor T. K. Peng, who has been giving me advices for my research and made the article possible. Professor Peng is currently a professor at College of Management of I-Shou University, Taiwan. His e-mail id is: tkpeng@isu.edu.tw.

REFERENCES

Bagchi, K.K., Udo, G.J., Kirs, P.J., Choden, K. (2015). Internet use and human values: Analyses of developing and developed countries. *Computers in Human Behavior, 50*, 76-90.
[http://dx.doi.org/10.1016/j.chb.2015.03.055]

Bannon, S., McGlynn, T., McKenzie, K., Quayle, E. (2015). The positive role of Internet use for young people with additional support needs: Identity and connectedness. *Computers in Human Behavior, 53*, 504-514.
[http://dx.doi.org/10.1016/j.chb.2014.11.099]

Blackwell, C.K., Lauricella, A.R., Conway, A., Wartella, E. (2014). Children and the Internet: Developmental implications of web site preferences among 8-to 12-year-old children. *Journal of Broadcasting and Electronic Media, 58*(1), 1-20.
[http://dx.doi.org/10.1080/08838151.2013.875022]

Boyd, d., Hargittai, E., Schultz, J., Palfrey, J. (2011). Why parents help their children lie to Facebook about age: Unintended consequences of the 'Children's Online Privacy Protection Act'. *First Monday, 16*(11)

Campbell, A.J., Cumming, S.R., Hughes, I. (2006). Internet use by the socially fearful: addiction or therapy? *Cyberpsychology and Behavior, 9*(1), 69-81.

[http://dx.doi.org/10.1089/cpb.2006.9.69] [PMID: 16497120]

Caplan, S.E. (2003). Preference for online social interaction: A theory of problematic Internet use and psychosocial well-being. *Communication Research, 30*(6), 625-648.
[http://dx.doi.org/10.1177/0093650203257842]

Chang, P.F., Choi, Y.H., Bazarova, N.N., Löckenhoff, C.E. (2015). Age Differences in Online Social Networking: Extending Socioemotional Selectivity Theory to Social Network Sites. *Journal of Broadcasting and Electronic Media, 59*(2), 221-239.
[http://dx.doi.org/10.1080/08838151.2015.1029126]

Consumer Reports. *CR Survey: 7.5 Million Facebook Users are Under the Age of 13, Violating the Site's Terms. Retrieved from: http://pressroom.consumerreports.org/pressroom/2011/05/cr-survey-75-million-facebook-users-are-under-the-age-of-13-violating-the-sites-terms-.html.*

Davis, R.A. (2001). A cognitive-behavioral model of pathological Internet use. *Computers in Human Behavior, 17*(2), 187-195.
[http://dx.doi.org/10.1016/S0747-5632(00)00041-8]

Ellison, N.B., Steinfield, C., Lampe, C. (2007). The benefits of Facebook "friends:" Social capital and college students' use of online social network sites. *Journal of Computer-Mediated Communication, 12*(4), 1143-1168.
[http://dx.doi.org/10.1111/j.1083-6101.2007.00367.x]

Facebook Report. *Facebook Reports Second Quarter 2015 Results. Retrieved from: http://investor.fb. com/releasedetail.cfm?ReleaseID=924562.*

Fung, H.H., Carstensen, L.L., Lang, F.R. (2001). Age-related patterns in social networks among European Americans and African Americans: implications for socioemotional selectivity across the life span. *The International Journal of Aging and Human Development, 52*(3), 185-206.
[http://dx.doi.org/10.2190/1ABL-9BE5-M0X2-LR9V] [PMID: 11407486]

Gerpott, T.J., Thomas, S. (2014). Empirical research on mobile Internet usage: A meta-analysis of the literature. *Telecommunications Policy, 38*(3), 291-310.
[http://dx.doi.org/10.1016/j.telpol.2013.10.003]

Griffiths, M. (2000). Does Internet and computer" addiction" exist? Some case study evidence. *Cyberpsychology and Behavior, 3*(2), 211-218.
[http://dx.doi.org/10.1089/109493100316067]

Internet World Stats. *Retrieved from: http://www.internetworldstats.com/.*

Kenski, K., Stroud, N.J. (2006). Connections between Internet use and political efficacy, knowledge, and participation. *Journal of Broadcasting and Electronic Media, 50*(2), 173-192.
[http://dx.doi.org/10.1207/s15506878jobem5002_1]

Kohlberg, L. (1984). *Essays on moral development* (Vol. 2). USA: Harper & Row.

Kohlberg, L., Kohlberg, L. (1969). Stage and Sequence: the Cognitive-Developmental Approach to Socialization. *Handbook of Socialization and Endash; Theory and research* (Vol. 2). Chicago: McNally.

Laconi, S., Tricard, N., Chabrol, H. (2015). Differences between specific and generalized problematic Internet uses according to gender, age, time spent online and psychopathological symptoms. *Computers in Human Behavior, 48*, 236-244.
[http://dx.doi.org/10.1016/j.chb.2015.02.006]

Lagraña, F. A. (2010). Ethical Issues Arising from the Usage of Electronic Communications in the Workplace. *Ethical Issues in E-Business: Models and Frameworks: Models and Frameworks,* 200.

Lee, Y.C. (2011). Competitive relationships between traditional and contemporary telecommunication services in Taiwan. *Telecommunication Policy, 35*(6), 543-554.
[http://dx.doi.org/10.1016/j.telpol.2011.04.017]

Lenhart, A. (2015). Teens, Technology and Friendships. *Pew Research Center. Retrieved from: http://www.pewinternet.org/2015/08/06/teens-technology-and-friendships/.*

Lin, W.Y., Zhang, X., Jung, J.Y., Kim, Y.C. (2013). From the wired to wireless generation? Investigating teens' Internet use through the mobile phone. *Telecommunication Policy, 37*(8), 651-661.
[http://dx.doi.org/10.1016/j.telpol.2012.09.008]

Livingstone, S., Haddon, L., Görzig, A., Ólafsson, K. Risks and safety on the internet: the perspective of European children: full findings and policy implications from the EU Kids Online survey of 9-16 year olds and their parents in 25 countries.

Lo, V.H., Wei, R. (2005). Exposure to Internet pornography and Taiwanese adolescents' sexual attitudes and behavior. *Journal of Broadcasting and Electronic Media, 49*(2), 221-237.
[http://dx.doi.org/10.1207/s15506878jobem4902_5]

Local Consumer Review Survey. *Retrieved from: www.brightlocal.com/wp-content/uploads/2013/06/Local-Consuner-Review-Survey-2013.pdf.*

Lu, X., Yeo, K.J. (2015). Pathological Internet use among Malaysia University Students: Risk factors and the role of cognitive distortion. *Computers in Human Behavior, 45*, 235-242.
[http://dx.doi.org/10.1016/j.chb.2014.12.021]

Lynch, M. (1994). *Ethical Issues in Electronic Information Systems.* The Geographer's Craft Project.

Ma, H.K. (2011). Internet addiction and antisocial internet behavior of adolescents. *ScientificWorldJournal, 11*, 2187-2196.
[http://dx.doi.org/10.1100/2011/308631] [PMID: 22125466]

Macer, D.R. (2008). *Moral games for teaching bioethics.* International Center for Health, Law and Ethics, Faculty of Law, University of Haifa.

Madden, M., Lenhart, A., Duggan, M., Cortesi, S., Gasser, U. (2013). *Teens and technology 2013.*

Washington, DC: Pew Internet & American Life Project.

Manago, A.M., Taylor, T., Greenfield, P.M. (2012). Me and my 400 friends: the anatomy of college students Facebook networks, their communication patterns, and well-being. *Developmental Psychology, 48*(2), 369-380.
[http://dx.doi.org/10.1037/a0026338] [PMID: 22288367]

Mou, Y., Lin, C.A. (2015). Exploring Podcast Adoption Intention *via* Perceived Social Norms, Interpersonal Communication, and Theory of Planned Behavior. *Journal of Broadcasting and Electronic Media, 59*(3), 475-493.
[http://dx.doi.org/10.1080/08838151.2015.1054997]

Nielsen. *Consumer Trust in Online, Social and Mobile Advertising Grows. Retrieved from: http://www.nielsen.com/us/en/insights/news/2012/consumer-trust-in-online-social-and-mobile-advertising-grows.html.*

Nikou, S., Bouwman, H. (2014). Ubiquitous use of mobile social network services. *Telematics and Informatics, 31*(3), 422-433.
[http://dx.doi.org/10.1016/j.tele.2013.11.002]

Rainiee, L. (2015). *Manners 2.0: Key findings about etiquette in the digital age.* Pew Research Center.

Renren. *Retrieved from: http://www.renren-inc.com/en/.*

Rideout, V.J., Foehr, U.G., Roberts, D.F. (2010). *Generation: Media in the Lives of 8-to 18-Year-Olds.* Menlo Park, CA: The Kaiser Family Foundation.

Shek, D.T., Tang, V.M., Lo, C.Y. (2008). Internet addiction in Chinese adolescents in Hong Kong: assessment, profiles, and psychosocial correlates. *ScientificWorldJournal, 8*, 776-787.
[http://dx.doi.org/10.1100/tsw.2008.104] [PMID: 18690381]

Southwell, B.G., Gilkerson, N.D., Depue, J.B., Shelton, A.K., Friedenberg, L.M., Koutstaal, W. (2010). Aging and the questionable validity of recognition-based exposure measurement. *Communication Research, 37*(5), 603-619.
[http://dx.doi.org/10.1177/0093650209356442]

Taipei Times. *Retrieved from: http://www.taipeitimes.com/News/main/history/2012/10/19.*

Taipei Times. *Retrieved from: http://www.taipeitimes.com/News/main/history/2014/09/01.*

Taiwan Network Information Center. *Retrieved from: http://www.seminar2014.twnic.tw/lecture.html.*

Tencent Announces. *First Quarter Results. Retrieved from: http://www.tencent.com/en-us/index.shtml.*

UDN News Database. *Retrieved from: http://udndata.com/.*

Tolbert, C.J., McNeal, R.S. (2003). Unraveling the effects of the Internet on political participation? *Political Research Quarterly, 56*(2), 175-185.

[http://dx.doi.org/10.1177/106591290305600206]

You, K.H., Lee, J.K., Kang, H., Go, E. (2015). Participatory or deliberative democracy? Exploring the mediation effects of perceived online deliberation and online interactive activities on social outcomes. *Telematics and Informatics, 32*(2), 205-214.
[http://dx.doi.org/10.1016/j.tele.2014.08.004]

Young, K.S. (1998). Internet addiction: The emergence of a new clinical disorder. *Cyberpsychology and Behavior, 1*(3), 237-244.
[http://dx.doi.org/10.1089/cpb.1998.1.237]

Zilberstein, K. (2013). Technology, Relationships and Culture: Clinical and Theoretical Implications. *Clinical Social Work Journal, 43*(2), 151-158.
[http://dx.doi.org/10.1007/s10615-013-0461-2]

SUBJECT INDEX

Kuo Hung Huang (Ed.)

D

E

F

G

L

M

N

O

P